Remote Sensing of the Environn
and Radiation Transfer

Anatoly Kuznetsov • Irina Melnikova •
Dmitry Pozdnyakov • Olga Seroukhova •
Alexander Vasilyev

Remote Sensing of the Environment and Radiation Transfer

An Introductory Survey

Springer

Anatoly Kuznetsov
Russian State
Hydrometeorological Univer
Atmospheric Physics
Malookhtinsky av. 98
195196 Sankt Petersburg
Russia
kuznetsov46@inbox.ru

Dmitry Pozdnyakov
Nansen International
Environmental and R
Satellite Remote Sensing for Aquatic Eco
14th Line, V.O. 7
199034 Sankt Petersburg
Russia
Dmitry.pozdnyakov@niersc.spb.ru

Alexander Vasilyev
St. Petersburg State University
Physics Department
Ul'yanovslkaya 1
198504 Sankt Petersburg
Petrodvorets
Russia
vsa@lich.phys.spbu.ru

Irina Melnikova
INENCO RAS
Atmospheric Physics
Kutuzova nab. 14
191187 Sankt Petersburg
Russia
Irina.Melnikova@pobox.spbu.ru

Olga Seroukhova
Russian State
Hydrometeorological Univer
Atmospheric Physics
Malookhtinsky av. 98
195196 Sankt Petersburg
Russia
serouhova@inbox.ru

ISBN 978-3-642-44116-5 ISBN 978-3-642-14899-6 (eBook)
DOI 10.1007/978-3-642-14899-6
Springer Heidelberg Dordrecht London New York

Printed on acid-free paper

Springer is part of Springer Science+Business Media (www.springer.com)

Contents

An Introduction to Remote Sensing

Practical work to senior undergraduate courses of lectures: "Remote sensing of the Environment" and "Radiative transfer in the atmosphere" by Anatoly Kuznetsov, Irina Melnikova, Olga Seroukhova, Dmitry Pozdnyakov, Alexander Vasilyev

Abstract

The interaction of the solar and heat radiation with the atmosphere and the earth's surface is the subject of this book. It is useful not only for students and professors but for wide circle of scientists involved in environmental studies.

The book contains descriptions of computer studying programs concerning different topics of courses on the remote sensing and radiative transfer. Thus it is not a textbook with a methodically presented subject of courses, but it includes only the basic ground for the comprehension of key topics of courses and provides the accomplishment of practical works using specially elaborated computer programs. The certain eclecticism of the book is explained with these circumstances.

We need to point out that practical works have been elaborated by authors independently that influences the style of different chapters and design of computer codes. The first chapter is mutual, Chaps. 2–9 is written by Anatoly Kuznetsov and Olga Seroukhova, part of the Chaps. 9, 11 and 12 – by Irina Melnikova, and Chaps. 10, 13–15 – by Alexander Vasilyev, and Chaps. 16–18 by Dmitry Pozdnyakov.

Themes of practical works reflect main sections of mentioned courses of lectures. The packet of computer programs is added to this book for the calculation of solar and heat radiation characteristics on the base of initial parameters of the atmosphere and surface (direct problems) and for retrieval optical parameters of the atmosphere and surface from radiation data (observed or simulated). Programs provide the dialog with the user and include graphics of results in some cases. For accomplishment of the practical work the student has to prepare initial data, to fulfill calculations with the corresponding program, to plot demanded graphics of calculated values dependencies on initial parameters and prepare the report with the use of the editors WORD and EXCEL. It promotes to deep and reliable comprehension of corresponding topics by students. It is important that all described approaches and computer programs are valuable resources for solving radiative transfer problems and they could be used by students for courses and diploma studies concerned calculation of radiative characteristics or solution of direct and inverse problems of remote sensing and radiative transfer.

Part I

The first chapter contains common information about radiative transfer in the atmosphere, which is necessary for understanding practical works.

The second chapter is devoted to specific features of black body and real body spectral brightness. Proposed tasks provide the understanding of conceptions of brightness temperature and emissivity of real bodies.

The third chapter considers the direct calculation of absorption coefficient of atmospheric gases with the use of parameters of the line structure of absorption bands. Results are necessary for the transmission function calculation.

The fourth chapter contains the calculation of transmission functions in different IR-spectral ranges on the base of different models of the absorption bands of the atmospheric gases. Results are used in following studies.

The fifth chapter includes the calculation of outgoing heat radiation of the Earth's surface and atmosphere for different cases of cloudy and clear atmosphere. Results allow demonstrating the separate contribution of the atmosphere and surface to the intensity of outgoing heat radiation.

The sixth chapter is about the construction and operation principles of the automatic one channel IR-radiometer.

In the seventh chapter it is proposed to accomplish the remote measurement of the temperature field of the surface with the automatic one channel IR-radiometer.

The eighth chapter is aimed to estimation of errors of the surface temperature retrieval, called by errors of measurement, a priori information and other factors in solving the inverse problem. Three different approaches are considered for estimating resulting uncertainties and their statistical characteristics.

The Chap. 9 contains two based methods of solving the inverse problem of the thermal remote sensing – retrieval of vertical temperature profiles from the spectral observations of IR intensity: the method of statistical regularization and the method of smoothing functional. Choosing different values of elements of temperature deviations matrix and the regularization parameter together with a priori information for removing the incorrectness allows demonstrating the inverse problem incorrectness and mathematical approaches.

The tenth chapter is dedicated to calculating optical characteristics of atmospheric aerosols. Different spectral dependencies of aerosol scattering and absorption coefficients are considered together with different shapes of phase functions, including big water droplets demonstrating rainbow maximums.

Part II

Chapters 11–14 consider different methods of the transfer theory for calculating radiative characteristics for models of cloudy and clear atmosphere: asymptotic formulas, Eddington formulas, Monte-Carlo approach, and single scattering approach. Specific features of every method are detailed outlined together with their exactness for different atmospheric models and geometry of illumination.

Chapter 15 contains peculiarities of anisotropic reflection from surface on the example of waved water surface. Different types of reflection form are considered. It is proposed to study the dependency of brightness coefficients on solar elevation and wind velocity and plot 3-D graphics for presenting brightness maximum and wideness of sunlight patches as functions of mentioned factors.

Chapter 16 considers the calculation and analysis of spectral distribution of the coefficient of diffuse reflection of solar radiation of the water column in deep and shallow basins.

Chapter 17 is devoted to calculation and analysis of variety of radiometric color characteristics of natural water applying to basins with high space heterogeneity of hydro-optical properties.

Chapter 18 contains the approach of the retrieval concentrations of optically active components of natural water from spectral distribution of the coefficient of diffuse reflection of solar radiation below water surface in problems of environment remote sensing.

The book contains the set of optical parameters corresponding to simple atmospheric and surface models, which could be useful for first approximation estimating of interaction of the Environments and electro-magnetic radiation.

Chapter 1
Radiation in the Earth Atmosphere

Abstract The first chapter contains general information about interaction of solar radiation with the atmosphere, elements of the radiative transfer theory, and base information about the scattering theory, which is necessary for understanding practical works. The basic notations of solar radiation characteristics and atmospheric optical parameters together with corresponding equations are presented.

1.1 Characteristics of the Radiation Field in the Atmosphere

In accordance with the contemporary conceptions, light (radiation) is an electromagnetic wave showing the quantum properties. Thus, strictly speaking, the processes of light propagation in the atmosphere should be described within the ranges of electrodynamics and quantum mechanics. Nevertheless, it is suitable to abstract from the electromagnetic nature of light to solve the number of problems (including the problems described in this book) and to consider radiation as an energy flux. Light characteristics governed by energy are called *the radiative characteristics*. This approach is usual for optics because the frequency of the electromagnetic waves within the optical ranges is huge and the receiver registers only energy, received during many wave periods (not a simultaneous value of the electromagnetic intensity).

The following main types of radiation (and their energy) are distinguished in solar radiation transferring throughout the atmosphere: *direct* radiation (radiation coming to the point directly from the Sun); *diffused solar* radiation (solar radiation scattering in the atmosphere); *reflected solar* radiation from surface; *self-atmospheric* radiation (*heat atmospheric* radiation) and *self-surface* radiation (*heat* radiation). The total combination of these radiations creates the *radiation field* in the Earth atmosphere, which is characterized with energy of radiation coming from different directions within different spectral ranges. As it is seen from above, it is possible to divide all radiation to the solar and self (heat) radiation. The maximum of the solar radiation is

I. Melnikova et al., *Remote Sensing of the Environment and Radiation Transfer*,
DOI 10.1007/978-3-642-14899-6_1, © Springer-Verlag Berlin Heidelberg 2012

Fig. 1.1 The intensity and the flux of radiation (radiance and irradiance)

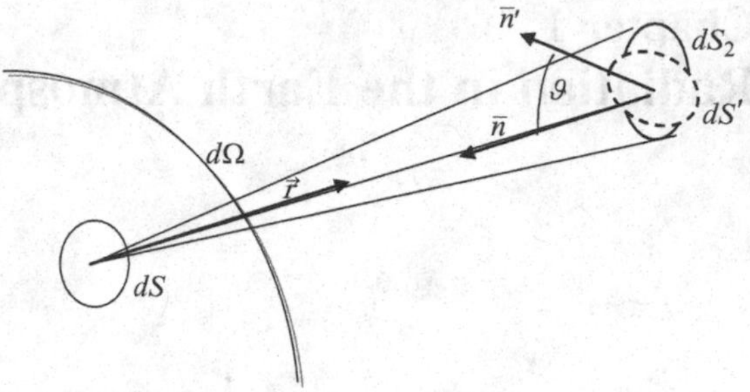

in the spectral ranges 0.3–1.0 μm which may be specified as *the short-wave spectral range*. Solar radiation integrated with respect to the wavelength over the considered spectral region will be called *total radiation*. Meanwhile, it should be noted that further definitions of the radiation characteristics are not linked within this limitation and could be used either for heat or for microwave ranges.

The notion of a monochromatic parallel beam (the plane electromagnetic wave of one concrete wavelength and one strict direction) is widely used in optics for the theoretical description of different processes. Usually solar radiation is set just in that form to describe its interactions with different objects. The principle of an independency of the monochromatic beams under their superposition is postulated, i.e. the interaction of the radiation beams coming from different directions with the object is considered as a sum of independent interactions along all directions. The physical base of the independency principle is an incoherence of the natural radiation sources.[1]

This standard operation is naturally used for the radiation field, i.e. the consideration of it as a sum of non-interacted parallel monochromatic beams. Furthermore, radiation energy can't be attributed to a single beam, because if energy were finite in the wavelength and direction intervals, it would be infinitesimal for the single wavelength and for the single direction. For characterizing radiation, it is necessary to pass from energy to its distribution over spectrum and directions.

Consider an emitting object (Fig. 1.1) implying not only the radiation source but also an object reflecting and scattering external radiation. Pick out a surface element dS, encircle the solid angle $d\Omega$ around the normal $\vec{r}$ to the surface. Then radiation energy would be proportional to the area dS, the solid angle $d\Omega$, as well as to the wavelength ranges $[\lambda, \lambda + d\lambda]$ and the time interval $[t, t + dt]$. The factor of the proportionality of radiation energy to the values dS, $d\Omega$, $d\lambda$ and dt would be

[1] It should be noted that monochromatic radiation is impossible in principle. It follows from the mathematical properties of the Fourier transformation: a spectrum consisting of one frequency is possible only with the time-infinite signal. Furthermore, the principle of the independency is not valid for the monochromatic beams because they always interfere. Both these contradictions are possible to remove if we are considering monochromatic radiation not as a physical but as a mathematical object, i.e. as a real radiation expansion into a sum (integral Fourier) of the harmonic terms. The separate item of this expansion is interpreted as monochromatic radiation.

specified *an intensity of the radiation* or *radiance* $I_\lambda(\vec{r}, t)$at the wavelength λ to the direction $\vec{r}$ at the moment t, namely:

$$I_\lambda(\vec{r}, t) = \frac{dE}{dSd\Omega d\lambda dt}. \tag{1.1}$$

In many cases we are interested not in energy emitted by the object but in energy of the radiation field, which is coming to the object (for example to the instrument input). Then it would be easy to convert the above specification of radiance. Consider the emitting object and set the second surface element of the equal area $dS_2 = dS$ at an arbitrary distance (Fig. 1.1). Let the system be situated in vacuum, i.e. radiation is not interacting during the way from dS to dS_2. Let the element dS_2 be perpendicular to the direction $\vec{r}$, then the solid angle at which the element dS_2 is seen from dS at the direction $\vec{r}$ is equal to the solid angle at which the element dS is seen from dS_2 at the opposite direction $(-\vec{r})$. The energies incoming to the surface elements dS and dS_2 are equal too thus, we are getting the consequence from the above definition of the intensity. The factor of the proportionality of emitted energy dE to the values dS, $d\Omega$, $d\lambda$ and dt is called an intensity (radiance) $I_\lambda(\vec{r}, t)$ incoming from the direction $\vec{r}$ to the surface element dS perpendicular to $\vec{r}$ at the wavelength λ at the time t i.e. Eq. 1.1. Point out the important demand of the perpendicularity of the element dS to the direction $\vec{r}$ in the definition of both the emitting and incoming intensity.

The definition of the intensity as a factor of the proportionality tends to have some formal character. Thus, the "physical" definition is often given: the intensity (radiance) is energy that incomes per unit time, per unit solid angle, per unit wavelength, per unit area perpendicular to the direction of incoming radiation, which has the units of watts per square meter per micron per steradian. This definition is correct if we specify energy to correspond not to the real unit scale (*sec*, *sterad*, *μm*, cm^2) but to the differential scale dt, $d\Omega$, $d\lambda$, dS, which is reduced then to the unit scale. Equation 1.1 is reflecting this obstacle.

Let the surface element dS', which radiation incomes to, be not perpendicular to the direction $\vec{r}$ but form the angle ϑ with it (Fig. 1.1). Specify *the incident angle* (the angle between the inverse direction $-\vec{r}$ and the normal to the surface) as $\vartheta = \angle(\vec{n}, -\vec{r})$. In that case we have to use the projection of the element dS' on a plane perpendicular to the direction of the radiation propagation in the capacity of the surface element dS, when defining the intensity as a factor of the proportionality. This projection is equal to $dS = dS' \cos\vartheta$. Then the expression $dE = I_\lambda(\vec{r}, t)dtd\lambda d\Omega dS' \cos\theta$ could be obtained from Eq. 1.1. It is suitable to attribute the sign to energy defined above. Actually, if we fix one concrete side of the surface dS' and assume the normal just to this side as a normal $\vec{n}$ then the angle ϑ varies from 0 to π, and the cosine from +1 to −1. Thus, incoming energy is positive and emitted energy is negative. It has transparent physical sense of the positive source and the negative sink of energy for the surface dS'. Now specify *the irradiance (the radiation flux of energy)* $F_\lambda(t)$ (often it is called the net spectral energy flux) as a factor of the proportionality of radiation energy dE'

incoming within a particular infinitesimal interval of wavelength $[\lambda, \lambda + d\lambda]$ and time $[t, t + dt]$ to the surface dS' from *the all directions* to values dt, $d\lambda$, dS' i.e.:

$$F_\lambda(t) = \frac{dE'}{dt d\lambda dS'}. \tag{1.2}$$

Adduce the "physical" definition of the irradiance that is often used instead of the "formal" one expressed by Eq. 1.2. Radiation energy incoming per unit area per unit time, per unit wavelength is called the radiation flux or irradiance. This definition corresponds correctly to Eq. 1.3 provided the meaning that energy is equivalent to the difference of incoming and emitted energy and uses the differential scale of area, time and wavelength. Proceeding from this interpretation, we will further use the term *energy* as a synonym of the *flux* implying the value of energy incoming per unit area, time and wavelength.

To characterize the direction of incoming radiation to the element dS' in addition to the angle ϑ, introduce the azimuth angle φ, which is counted off as an angle between the projection of the vector $\vec{r}$ to the plane dS and any direction on this plane $(0 \le \varphi \le 2\pi)$. Actually we are using the spherical coordinates system. Energy dE' incoming to the surface dS' from all directions is expressed in terms of energy from a concrete direction $dE(\vartheta,\varphi)$ as: $dE' = \int\limits_{\Omega=4\pi} dE(\vartheta, \varphi) d\Omega$, where the integration is accomplished over the whole sphere. Using the well-known expression for an element of the solid angle in the spherical coordinates $d\Omega = d\varphi \sin\vartheta d\vartheta$ we will get $dE' = \int\limits_0^{2\pi} d\varphi \int\limits_0^{\pi} dE(\vartheta, \varphi) \sin\vartheta d\vartheta$.

After the substituting this expression to (1.2) we will get the formula to express the irradiance:

$$F_\lambda(t) = \int\limits_0^{2\pi} d\varphi \int\limits_0^{\pi} I_\lambda(\vartheta, \varphi, t) \cos\vartheta \sin\vartheta d\vartheta \tag{1.3}$$

In addition to direction (ϑ, φ), wavelength λ and time t the solar radiance in the atmosphere depends on placement of the element dS. Owing to the sphericity of the Earth and its atmosphere, it is convenient to put the position of this element in the spherical coordinate system with its beginning in the Earth center. Nevertheless, taking into account that the thickness of the atmosphere is much less than the Earth radius is, in the number of problems the atmosphere could be considered by convention as a plane limited with two infinite boundaries: the bottom – a ground surface and the top – a level, above which the interaction between radiation and atmosphere could be neglected. Further, we are considering only *the plane-parallel atmosphere approximation*. Then the position of the element dS could be characterized with Cartesian coordinates (x, y, z) choosing the altitude as axe z (to put z axis perpendicular to the top and bottom planes from the bottom to the top). Thus, in general case the radiance in the atmosphere could be written as $I_\lambda(x, y, z, \vartheta, \varphi, t)$. Under the natural radiation sources (in particular – the solar one) we could neglect the behavior of the radiance in the time domain comparing with the time scales considered in the concrete problems (e.g. comparing with the instrument registration time). The radiation field under such conditions is called a *stationary* one. Further, it is possible to ignore the influence of the horizontal heterogeneity of the atmosphere on the radiation field comparing with the vertical one, i.e. don't consider the dependence of the radiance upon axes x and y. This radiation filed is called a *horizontally homogeneous* one. Further, we are considering only stationary and horizontally homogeneous radiation fields. Besides, following the traditions the subscript λ is omitted at the monochromatic values if the obvious wavelength dependence is not mentioned.

It is naturally to count off the angle ϑ from the selected direction z in the atmosphere. This angle is called zenith incident angle (it characterizes the inclination of incident radiation from zenith). The angle ϑ is equal to zero if radiation comes from zenith, and it is equal to π if radiation comes from nadir. As before we are counting off the azimuth angle from an arbitrary direction on the plane, parallel to the boundaries of the atmosphere. Then the integral (1.3) could be written as a sum of two integrals: over upper and lower hemisphere:

$$F(z) = F^{\downarrow}(z) + F^{\uparrow}(z),$$

$$F^{\downarrow}(z) = \int_0^{2\pi} d\varphi \int_0^{\pi/2} I(z,\vartheta,\varphi)\cos\vartheta\sin\vartheta d\vartheta, \tag{1.4}$$

$$F^{\uparrow}(z) = \int_0^{2\pi} d\varphi \int_{\pi/2}^{\pi} I(z,\vartheta,\varphi)\cos\vartheta\sin\vartheta d\vartheta.$$

The value $F^{\downarrow}(z)$ is called a *downward flux* (*downwelling irradiance*), the value $F^{\uparrow}(z)$ – an *upward flux* (*upwelling irradiance*), both are also called *semispherical fluxes* expressed in watts per square meter (per micron). The physical sense of these definitions is evident. The downward flux is radiation energy passing through the level z down to the ground surface and the upward flux is energy passing up from the ground surface. The downward flux is always positive ($\cos\vartheta > 0$), upward is always negative ($\cos\vartheta < 0$). In practice (for example during measurements) it is advisable to consider both fluxes as positive ones. We will follow this tradition. Then for the upward flux in Eq. 1.4 the value of $\cos\vartheta$ is to be taken in magnitude, and the total flux will be equal to the difference of the semispherical fluxes $F(z) = F^{\downarrow}(z) - F^{\uparrow}(z)$. This value is often called a *(spectral) net radiant flux* expressed in watts per square meter (per micron).

Consider two levels in the atmosphere, defined by the altitudes z_1 and z_2 (Fig. 1.2). Obtain solar radiation energy $R(z_1, z_2)$ (per units area, time and wavelength) absorbed by the atmosphere between these levels. Manifestly, it is necessary to subtract outcoming energy from the incoming:

Fig. 1.2 The net radiant flux

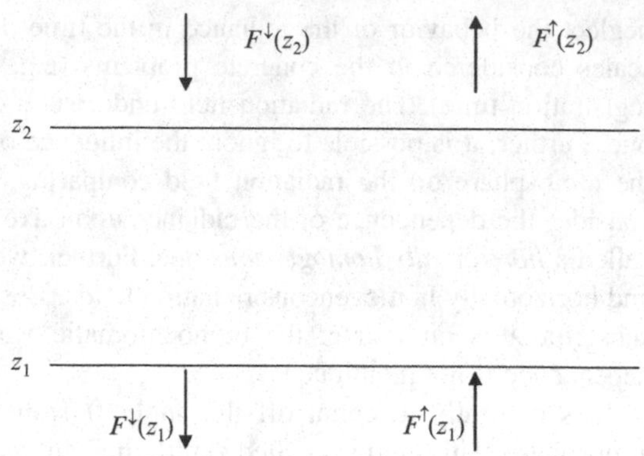

$$R(z_1,z_2) = F^{\downarrow}(z_2) + F^{\uparrow}(z_1) - F^{\downarrow}(z_1) - F^{\uparrow}(z_2) = F(z_2) - F(z_1) \qquad (1.5)$$

The value $B(z_1, z_2)$ is called a *radiative divergence in the layer between levels z_1 and z_2*. It is extremely important value for studying atmospheric energetics because it determines the warming of the atmosphere, and it is also important for studying the atmospheric composition because the spectral dependence of $R(z_1, z_2)$ allows to estimate the type and the content of specific absorbing materials (atmospheric gases and aerosols) within the layer in question. Hence, the values of the semispherical fluxes determining the radiative divergence are also of greatest importance for the mentioned class of problems.

Incident solar radiation incoming to the top of the atmosphere is practically always considered as one-directional radiation in the problems in question. Actually, it is possible to neglect the angular spread of the solar beam because of the infinitesimal radiuses of the Earth and the Sun comparing with the distance between them. Thus, we are considering the case of the plane parallel horizontally homogeneous atmosphere illuminated by a parallel solar beam. Some difficulties are appearing during the application of the above definitions to this case because we must attribute certain energy to the one-directional beam.

The radiance definition corresponding to Eq. 1.1 is not applicable in this case because it does not show the dependence of energy dE upon solid angle $d\Omega$ (formally following Eq. 1.1 we would get the zero intensity). As for the irradiance definition (1.3), it is applicable. Thus, it makes sense to examine the very irradiance of the strictly one-directional beams. Then the dependence of energy dE' upon the area of the surfaces dS' projection in Eq. 1.3 appears for differently orientated surfaces dS', which gives the following:

$$F(\vartheta) = F_0 \cos\vartheta, \qquad (1.6)$$

where F_0 is the irradiance for the perpendicular incident beam, $F(\vartheta)$ is the irradiance for the incident angle ϑ.

The incident flux F_0 is of fundamental importance for atmospheric optics and energetics. This flux is radiation energy incoming to the top of the atmosphere per unit area, per unit intervals of the wavelength and time in case of the average

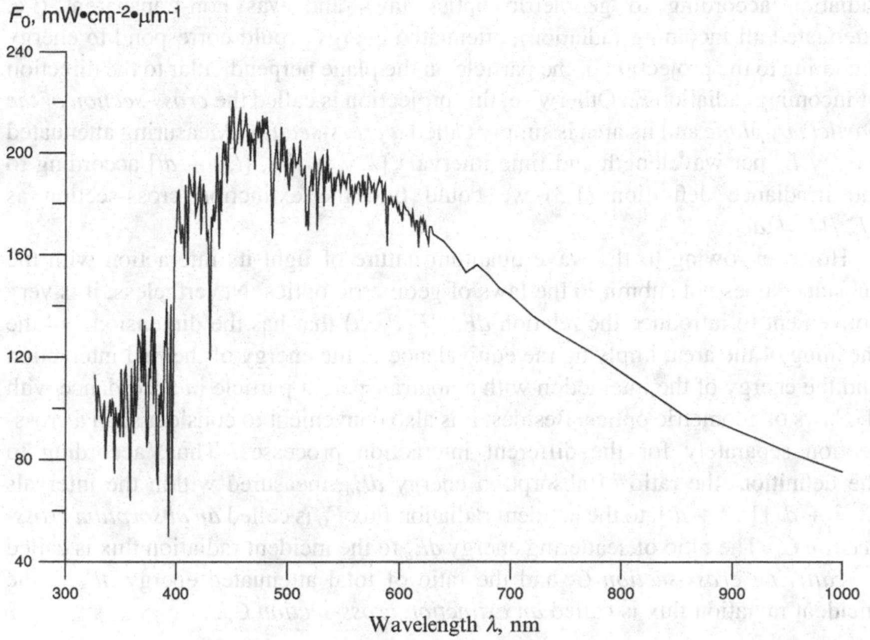

Fig. 1.3 Spectral extraterrestrial solar flux

Fig. 1.4 Definition of the cross-section of the interaction

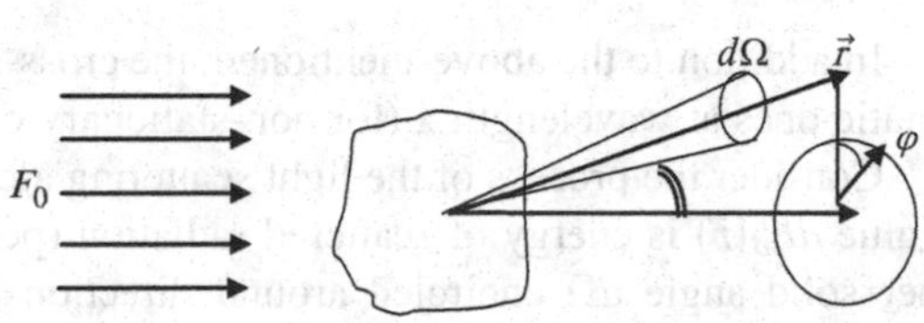

distance between the Sun and the Earth, and it is called *a spectral solar constant.* Figure 1.3 illustrates the *solar constant* F_0 as a function of wavelength.

1.2 Interaction of the Radiation and Atmosphere

Consider a symbolic particle (a gas molecule, an aerosol particle) that is illuminated by the parallel beam F_0 (Fig. 1.4). The process of the interaction of radiation and this particle is assembled of *the radiation scattering* on the particle and of *the radiation absorption* by the particle. Together these processes constitute *the radiation extinction* (the irradiance after interaction with the particle is attenuated by the processes of scattering and absorption along the incident beam direction $\vec{r}_0$). Let absorbed energy be equal to E_a, scattered in all directions energy be equal to E_s, and total attenuated energy be equal to $E_e = E_a + E_s$. If the particle interacted with radiation according to geometric optics laws and was non-transparent (i.e. attenuated all incoming radiation), attenuated energy would correspond to energy incoming to the projection of the particle on the plane perpendicular to the direction of incoming radiation $\vec{r}_0$. Otherwise, this projection is called the *cross-section of the particle by plane* and its area is simply called *a cross-section*. Measuring attenuated energy E_a per wavelength and time intervals $[\lambda, \lambda + d\lambda]$, $[t, t + dt]$ according to the irradiance definition (1.3) we could find the extinction cross-section as $dE_e/(F_0 d\lambda dt)$.

However, owing to the wave quantum nature of light its interaction with the substance does not submit to the laws of geometric optics. Nevertheless, it is very convenient to introduce the relation $dE_e/(F_0 d\lambda dt)$ that has the dimension and the meaning of the area, implying the equivalence of the energy of the real interaction and the energy of the interaction with a nontransparent particle in accordance with the laws of geometric optics. Besides, it is also convenient to consider such a cross-section separately for the different interaction processes. Thus, according to the definition, the ratio of absorption energy dE_a, measured within the intervals $[\lambda, \lambda + d\lambda]$ $[t, t + dt]$, to the incident radiation flux F_0 is called *an absorption cross-section* C_a. The ratio of scattering energy dE_s to the incident radiation flux is called *a scattering cross-section* C_s and the ratio of total attenuated energy dE_s to the incident radiation flux is called *an extinction cross-section* C_e:

$$C_a = \frac{dE_a}{F_0 d\lambda dt}, \quad C_s = \frac{dE_s}{F_0 d\lambda dt}, \quad C_e = \frac{dE_e}{F_0 d\lambda dt} = C_a + C_s \tag{1.7}$$

In addition to the above-mentioned, the cross-sections are defined as monochromatic ones at wavelength λ (for non-stationary case – at time t as well).

Consider the process of the light scattering along direction $\vec{r}$ (Fig. 1.4). Here the value $dE_d(\vec{r})$ is energy of scattered radiation (per intervals $[\lambda, \lambda + d\lambda]$ $[t, t + dt]$) per solid angle $d\Omega$ encircled around direction $\vec{r}$. Define *the directed scattering cross-section* analogously to the scattering cross-section expressed by Eq. 1.6.

$$C_d(\vec{r}) = \frac{dE_d(\vec{r})}{F_0 d\lambda dt d\Omega}. \tag{1.8}$$

Wavelength λ and time t are corresponding to the cross-section $C_d(\vec{r})$.

Total scattering energy is equal to the integral from $dE_d(\vec{r})$over all directions $dE_s = \int_{4\pi} dE_d d\Omega$. The link between the cross-sections of scattering and directed scattering is defined as $C_s = \int_{4\pi} C_d d\Omega$.

After passing to a spherical coordinate system and introducing two parameters: *the scattering angle* γ defined as an angle between directions of the incident and scattered radiation ($\gamma = \angle(\vec{r}_0, \vec{r})$) and *the scattering azimuth* φ we obtain $C_s = \int_0^{2\pi} d\phi \int_0^{\pi} C_d(\gamma, \varphi) \sin\gamma d\gamma$.

The directed scattering cross-section $C_d(\gamma,\varphi)$according to its definition could be treated as follows: as the value $C_d(\gamma,\varphi)$is higher, then light scatters stronger to the very direction (γ, φ) comparing to other directions. It is necessary to pass to a dimensionless value for comparison of the different particles using the directed scattering cross-section. For that the value $C_d(\gamma, \varphi)$ has to be normalized to the integral C_s and the result has to be multiplied by a solid angle. The resulting characteristic is called *a phase function* and specified with the following relation:

$$x(\gamma,\varphi) = 4\pi\frac{C_d(\gamma,\varphi)}{C_s}. \tag{1.9}$$

The substitution of the value $C_d(\gamma, \varphi)$ from Eq. 1.8 to Eq. 1.9 gives *the phase function normalization*:

$$\frac{1}{4\pi}\int_0^{2\pi} d\varphi \int_0^{\pi} x(\gamma,\varphi)\sin\gamma d\gamma = 1. \tag{1.10}$$

If the scattering is equal over all directions, i.e. $C_d(\gamma, \varphi) = const$, it is called *isotropic* and the relation $x(\gamma, \varphi) \equiv 1$ follows from the normalization (1.10). Thus, the multiplier 4π is used in Eq. 1.9 for convenience. In many cases, (for example the molecular scattering, the scattering on spherical aerosol particles) the phase function does not depend on the scattering azimuth. Further, we are considering only such phase functions. The integral from the phase function in limits between zero and scattering angle γ $\frac{1}{2}\int_0^{\gamma} x(\gamma)\sin\gamma d\gamma$ could be interpreted as *a probability of scattering to the angle interval* [0, γ]. It is easy to test this integral for satisfying all demands of the notion of the "probability". Hence the phase function $x(\gamma)$ is *the probability density of radiation scattering to the angle* γ. Often this assertion is accepted as a definition of the phase function.

The real atmosphere contains different particles interacting with solar radiation: gas molecules, aerosols particles of different size, shape and chemical composition, and cloud droplets. Therefore, we are interested in the interaction not with the separate particles but with a total combination of them. In the theory of radiative transfer and in atmospheric optics it is usual to abstract from the interaction with a separate particle and to consider the atmosphere as a continuous medium for simplifying the description of the interaction between solar radiation and all atmospheric components. It is possible to attribute the special characteristics of the interaction between the atmosphere and radiation to an elementary volume (formally infinitesimal) of this continuous medium.

Scrutinize *the elementary volume* of this continuous medium $dV = dSdl$ (Fig. 1.5), on which parallel flux of solar radiation F_0 incomes normally to the side dS. The interaction of radiation and elementary volume is reduced to the processes of scattering, absorption and radiation extenuation after radiation transfers through the elementary volume. Specify the radiation flux as $F = F_0 - dF$

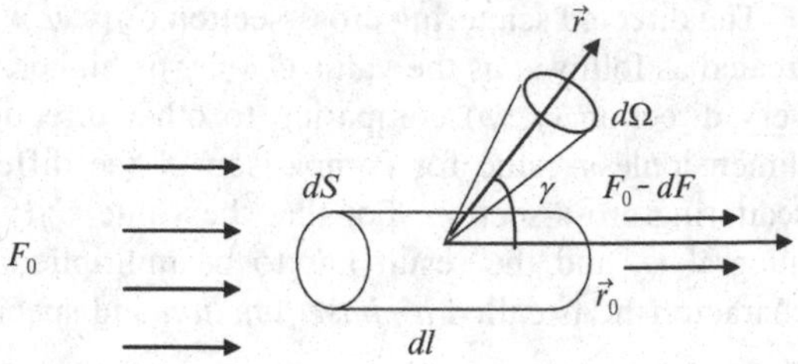

Fig. 1.5 Interaction between radiation and elementary volume of the scattering medium

after its penetrating the elementary volume (along the incident direction $\vec{r}_0$). Take the relative change of incident energy as an extinction characteristic $\frac{dE_e}{E_0} = \frac{(F_0-F)dSd\lambda dt}{F_0 dSd\lambda dt} = \frac{dF}{F_0}$.

As it is manifestly proportional to the length dl in the extenuating medium, then it is possible to take *the volume extinction coefficient* α as a characteristic of radiation, attenuated by the elementary volume. This coefficient is equal to a relative change of incident energy (measured in intervals $[\lambda, \lambda + d\lambda]$, $[t, t + dt]$) normalized to the length dl (i.e. reduced to the unit length) according to the definition $\alpha = \frac{dE_e}{E_0 dl} = \frac{dF}{F_0 dl}$. The analogous definitions of *the volume scattering* σ *and absorption* κ *coefficients* follow from the equality of extinction energy and the sum of the scattering and absorption energies.[2] $\sigma = \frac{dE_s}{E_0 dl}$, $\kappa = \frac{dE_a}{E_0 dl}$, $a = \sigma + k$.

Let us link the characteristics of the interaction between radiation and a separate particle with the elementary volume. If every particle interacts with radiation independently of others, then extinction energy of the elementary volume is equal to a sum of extinction energies of all particles in the volume. Suppose that all particles are similar; they have an extinction cross-section C_e, their number concentration (number of particle in the unit volume) is equal to n, and the particle number in the elementary volume is ndV. Then for the extinction coefficient we are obtaining the relation $\alpha = \frac{ndVC_e F_0 d\lambda dt}{F_0 dSd\lambda dtdl} = nC_e$. Thus, the volume extinction coefficient is equal to the product of particle number concentration by the extinction cross-section of one particle.[3]

If there are extenuating particles of M kinds with concentrations n_i and cross-sections $C_{e,I}$ in the elementary volume of the medium then it is valid: $dE_e = \sum_{i=1}^{M} n_i dVC_{e,i}F_0 d\lambda dt$. Analogously considering the energies of scattering, absorption and directed scattering, we are obtaining the formulas, which link the volume coefficients and cross-sections of the interaction:

[2] Notice, that introduced volume coefficients have dimension of the inverse length (m^{-1}, km^{-1}) and such values usually called "linear" not "volume". Further, we will substantiate this terminological contradiction.

[3] Just by this reason, the term "volume" and not "linear" is used for the coefficient. It is defined by numerical concentration in the unit volume of the air.

$$\alpha = \sum_{i=1}^{M} n_i C_{e,i}, \quad \sigma = \sum_{i=1}^{M} n_i C_{s,i}, \quad \kappa = \sum_{i=1}^{M} n_i C_{a,i}, \quad \sigma x(\gamma) = \sum_{i=1}^{M} n_i C_{s,i} x_i(\gamma)\,. \tag{1.11}$$

Point out that the separate items make sense of the volume coefficients of the interaction for the separate kinds of particles. Therefore, highly important for the practical problems the "summation rules" are following from Eq. 1.11. These rules allow deriving separately coefficients of the interaction and the phase function for each from M components and then to calculate the total characteristics of the elementary volume with the formulas:

$$\alpha = \sum_{i=1}^{M} \alpha_i, \quad \sigma = \sum_{i=1}^{M} \sigma_i, \quad \kappa = \sum_{i=1}^{M} \kappa_i, \quad x(\gamma) = \sum_{i=1}^{M} \sigma_i x_i(\gamma) \Big/ \sum_{i=1}^{M} \sigma_i. \tag{1.12}$$

These rules also allow calculating characteristics of the molecular and aerosol scattering and absorption of radiation in the atmosphere separately. Then Eq. 1.12 are transforming to the following:

$$\alpha = \sigma_m + \sigma_a + \kappa_m + \kappa_a, \quad \sigma = \sigma_m + \sigma_a \, \kappa = \kappa_m + \kappa_a,$$
$$x(\gamma) = \frac{\sigma_m x_m(\gamma) + \sigma_a x_a(\gamma)}{\sigma_m + \sigma_a}. \tag{1.13}$$

where σ_m, κ_m, $x_m(\gamma)$ are the volume coefficients of the molecular scattering, absorption and molecular phase function for the atmospheric gases respectively and σ_a, κ_a, $x_a(\gamma)$ are the analogous aerosol characteristics.

The volume coefficient and the phase function of the molecular scattering are expressed as follows:

$$\sigma_m = \frac{8}{3}\pi^3 \frac{(m^2-1)^2}{n\lambda^4} \frac{6+3\delta}{6-7\delta}, \quad x_m(\gamma) = \frac{3}{4+2\delta}(1+\delta+(1-\delta)\cos^2\gamma), \tag{1.14}$$

where m is the refractive index of the air, n is the number concentration of the air molecules, λ is the radiation wavelength, δ is the depolarization factor (for the air it is equal $\delta = 0.0279$).

The calculations of the aerosol scattering and absorption cross-sections so as an aerosol phase function are based on the simulations. The aerosol particles are approximated with the certain geometrical solids of the known chemical composition. Usually there are considered the homogeneous spherical particles. The calculation of the optical characteristics for such particles is accomplished according to the formulas of Mie theory, which we are not adducing here referring the reader to corresponding books.

The phase function of the aerosol scattering is presented in the above-mentioned calculations as a look-up table with the grid over the scattering angle. It is not

convenient for some problems where the phase function needs an analytical approximation. The one of the widely used approximations is a Henyey-Greenstein phase function:

$$x(\gamma) = \frac{1 - g^2}{(1 + g^2 - 2g\cos\gamma)^{3/2}}, \tag{1.15}$$

where g is the approximation parameter ($0 \leq g < 1$),

$$g = \frac{1}{3}x_1 = \frac{1}{4\pi}\int_0^{2\pi} d\varphi \int_{-1}^{1} x(\gamma)\gamma d\gamma, \tag{1.16}$$

it coincides with the mean cosine of the scattering angle, changes in the ranges [0,1], and is called often *the asymmetry factor* because it governs the degree of the phase function forward extension.

The function describes the main property of the aerosol phase functions – the forward peak – (the prevalence of the scattering to the forward hemisphere $0 \leq \gamma \leq \pi/2$ over the scattering to the back hemisphere $\pi/2 \leq \gamma \leq \pi$) and it is very suitable for the theoretical consideration, as it will be shown further.

1.3 Radiative Transfer in the Atmosphere

Within the elementary volume, the enhancing of energy along the length dl could occur in addition to the extinction of the radiation considered above. Heat radiation of the atmosphere within the infrared range is an evident example of this process, though as it will be shown further the accounting of energy enhancing is really important in the short-wave range either. Value dE – the enhancing of energy – is proportional to the spectral $d\lambda$ and time dt intervals, to the arc of solid angle $d\Omega$ encircled around the incident direction and to the value of emitting volume $dV = dSdl$. Specify the *volume emission coefficient* ε as a coefficient of this proportionality $\varepsilon = \frac{dE_r}{dVd\Omega d\lambda dt}$.

Consider now the elementary volume of medium within the radiation field. In general case both the extinction and the enhancing of energy of radiation passing through this volume are taking place (Fig. 1.6). Let I be the radiance incoming to the volume perpendicular to the side dS and $I + dI$ be the radiance after passing the volume along the same direction. According to energy definition in Eq. 1.1 incoming energy is equal to $E_0 = IdSd\Omega d\lambda dt$ then the change of energy after passing the volume is equal to $dE = dIdSd\Omega d\lambda dt$. According to the law of the conservation of energy, this change is equal to the difference between enhancing dE_r and extincting dE_e energies. Then, taking into account the above definitions of the volume

emission and extinction coefficients, we are defining *the radiative transfer equation*:

$$\frac{dI}{dl} = -\alpha I + \varepsilon \tag{1.17}$$

In spite of the simple form, Eq. 1.17 is the general transfer equation that accepts the coefficients α and ε as variable values. This derivation of the radiative transfer equation is phenomenological. The rigorous derivation must be done using the Maxwell equations.

Move to the consideration of particular cases of transfer Eq. 1.17 in conformity with *shortwave solar radiation in the Earth atmosphere*. Within the shortwave spectral range we omit the heat atmospheric radiation against the solar one and seem to have the relation $\varepsilon = 0$. However, we are taking into account that the enhancing of emitted energy within the elementary volume could occur also owing to the scattering of external radiation coming to the volume along the direction of the transfer in Eq. 1.17 (i.e. along the direction normal to the side *dS*). Specify this direction $\vec{r}_0$ and scrutinize radiation scattering from direction $\vec{r}$ with scattering angle γ (Fig. 1.6). Encircling the similar volume around direction $\vec{r}$(it is denoted as a dashed line), we are obtaining energy scattered to direction $\vec{r}_0$. Then employing precedent value of energy E_0, we are obtaining the contribution to the emission coefficient corresponded to direction $\vec{r}$:

$$d\varepsilon(\vec{r}) = \frac{\frac{\sigma}{4\pi}x(\gamma)I(\vec{r})dSd\Omega d\lambda dtd\Omega dl}{dVd\Omega d\lambda dt} = \frac{\sigma}{4\pi}x(\gamma)I(\vec{r})d\Omega.$$

Then it is necessary to integrate value $d\varepsilon(\vec{r})$ over all directions and it leads to *the integro-differential transfer equation while taking into account the scattering*:

$$\frac{dI(\vec{r}_0)}{dl} = -\alpha I(\vec{r}_0) + \frac{\sigma}{4\pi}\int_{4\pi} x(\gamma)I(\vec{r})d\Omega. \tag{1.18}$$

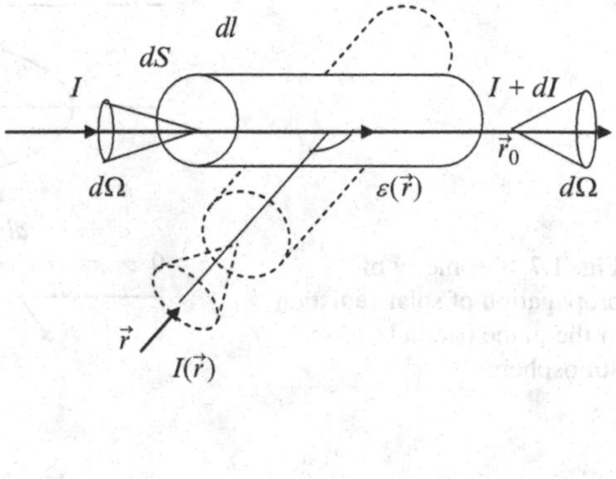

Fig. 1.6 The derivation of the radiative transfer equation

Consider the geometry of solar radiation spreading throughout the atmosphere for concretisation Eq. 1.18 as Fig. 1.7 illustrates. We are presenting the atmosphere as a model of the plane-parallel and horizontally homogeneous layer. The direction of the radiation spreading is characterized with the zenith angle ϑ and with the azimuth φ counted off an arbitrary direction at a horizontal plane. Set all coefficients in Eq. 1.18 depending on the altitude (it is completely corresponded to reality).

Length element dl in the plane-parallel atmosphere is $dl = -dz/\cos\vartheta$. The ground surface at the bottom of the atmosphere is neglected for the present (i.e. it is accounted that the radiation incoming to the bottom of the atmosphere is not reflected back to the atmosphere and it is equivalent to the almost absorbing surface). Within this horizontally homogeneous medium, the radiation field is also the horizontally homogeneous owing to the shift symmetry (the invariance of all conditions of the problem relatively to any horizontal displacement). Thus, the radiance is a function of only three coordinates: altitude z and two angles, defining direction (ϑ, φ). Hence, Eq. 1.18 could be written as:

$$\frac{dI(z,\vartheta,\varphi)}{dz}\cos\vartheta = \alpha(z)I(z,\vartheta,\varphi) - \frac{\sigma(z)}{4\pi}\int\limits_0^{2\pi} d\varphi' \int\limits_0^{\pi} x(z,\gamma)I(z,\vartheta',\varphi')\sin\vartheta' d\vartheta' \tag{1.19}$$

where scattering angle γ is an angle between directions (ϑ, φ) and $(\vartheta'\varphi')$. It is easy to express the scattering angle through ϑ, φ: to consider the scalar product of the orts in Cartesian coordinate system and then pass to the spherical coordinates. This procedure yields the following relation known as Cosine law for the spheroid triangles $\cos\gamma = \cos\vartheta\,\cos\vartheta' + \sin\vartheta\,\sin\vartheta'\cos(\varphi - \varphi')$.

To begin with, consider the simplest particular case of transfer Eq. 1.18. Let us neglect the radiation scattering i.e. the term with the integral. For atmospheric

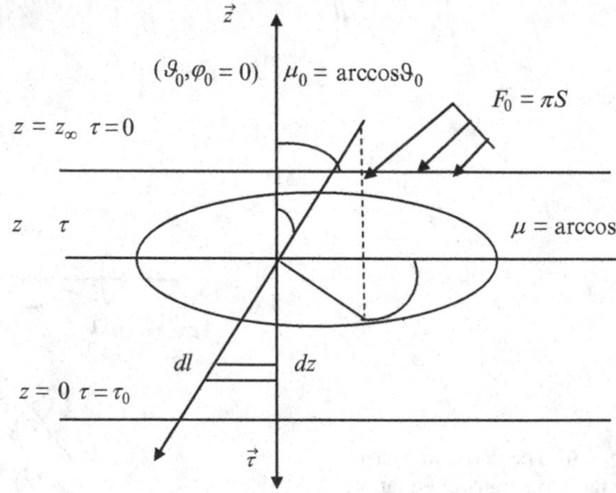

Fig. 1.7 Geometry of propagation of solar radiation in the plane parallel atmosphere

optics, it conforms to the direction of the direct radiation spreading (ϑ_0, φ_0). Actually in the cloudless atmosphere, the intensity of solar direct radiation is essentially greater than the intensity of scattered radiation. In this case, the direction of solar radiation is only one, the intensity depends only on the altitude, and the transfer Eq. 1.19 transforms to the following $\frac{dI(z)}{dz}\cos\vartheta_0 = \alpha(z)I(z)$ and it is always $\sin\vartheta_0 > 0$ here. Differential equation together with boundary condition $I = I(z_\infty)$, where z_∞ is the altitude of the top of the atmosphere (the level above which it is possible to neglect the interaction between solar radiation and atmosphere) is elementary solved that leads to:

$$I(z) = I(z_\infty)\exp\left(-\frac{1}{\cos\vartheta_0}\int_z^{z_\infty}\alpha(z')dz'\right). \quad (1.20)$$

This relation illustrates the exponential decrease of the intensity in the extinct medium and it is called Beer's law.

Introduce the dimensionless value:

$$\tau(z) = \int_z^{z_\infty}\alpha(z')dz', \quad (1.21)$$

that is called *the optical depth* of the atmosphere at altitude z. Its important particular case is *the optical thickness* of the atmosphere in whole $\tau_0 = \int_0^{z_\infty}\alpha(z')dz'$.

Then Beer's law is written as:

$$I(z) = I(z_\infty)\exp(-\tau(z)/\cos\vartheta_0). \quad (1.22)$$

As it follows from definitions (1.20) and (1.22) and from "summation rules" (1.12), the analogous rules are correct for the optical deepness and for the optical thickness: $\tau(z) = \sum_{i=1}^{M}\tau_i(z)$, $\tau_0 = \sum_{i=1}^{M}\tau_{0,i}$.

Therefore, it is possible to specify the optical thickness of the molecular scattering, the optical thickness of the aerosol absorption etc.

According to the accepted in Sect. 1.1 condition we are considering solar radiation incoming to the plane atmosphere top as an incident solar parallel flux F_0 from direction (ϑ_0, φ_0). Then, deducing the intensity through delta-function (1.10) and substituting it to the formula of the link between the flux and intensity (1.5) it is possible to obtain Beer's Law for the solar irradiance incoming to the horizontal surface at the level τ:

$$F_d(z) = F_0\cos\vartheta_0\exp(-\tau(z)/\cos\vartheta_0). \quad (1.23)$$

It is to be pointed out that the function $P(\tau, \vartheta) = \exp(-\tau_0/\cos\vartheta_0)$ is called the *transmission function* and used in the calculation of the heat radiation.

Return to the general case of the transfer equation and taking into account scattering (1.19). Accomplish the transformation to the dimensionless parameters in the transfer equation for convenience of further analysis. In accordance with optical thickness definition (1.21) the function $\tau(z)$ is monotonically decreasing with altitude that follows from condition $\alpha(z') > 0$. In this case there is an inverse function $z(\tau)$ that is also decreasing monotonically. Using the formal substitution of function $z(\tau)$ rewrite the transfer equation and pass from vertical coordinate τ to coordinate z, moreover, the boundary condition is at the top of the atmosphere $\tau = 0$ and at the bottom $\tau = \tau_0$, and the direction of axis τ is opposite to axis z. It follows from the definition (1.21): $d\tau = -\alpha(z)dz$. Specify $\mu = \cos\vartheta$ and pass from the zenith angle to its cosine (the formal substitution $\vartheta = \arccos\mu$ with taking into account $\sin\vartheta d\vartheta = -d\mu$). Finally, divide both parts of the equation to value $\alpha(\tau)$, and obtain instead Eq. 1.19 the following equation:

$$\mu\frac{dI(\tau,\mu,\varphi)}{d\tau} = -I(\tau,\mu,\varphi) + \frac{\omega_0(\tau)}{4\pi}\int_0^{2\pi} d\varphi' \int_{-1}^{1} x(\tau,\gamma)I(\tau,\mu',\varphi')d\mu', \qquad (1.24)$$

where

$\omega_0(\tau) = \frac{\sigma(\tau)}{\alpha(\tau)} = \frac{\sigma(\tau)}{\sigma(\tau)+\kappa(\tau)}$, and the scattering angle cosine $\cos\gamma = \mu\mu' + \sqrt{1-\mu^2}$ $\sqrt{1-\mu'^2}\cos(\varphi-\varphi')$.

Dimensionless value ω_0 is called *the single scattering albedo* or otherwise *the probability of the quantum surviving* per the single scattering event. If there is no absorption ($\kappa = 0$) then the case is called *conservative scattering*, $\omega_0 = 1$. If the scattering is absent then the extinction is caused only by absorption, $\sigma = 0$, $\omega_0 = 0$ and the solution of the transfer equation is reduced to Beer's law. After consideration of these cases, the sense of value ω_0 is following: it defines the part of scattered radiation relatively to the total extinction, and corresponds to the probability of the quantum to survive and accepts the quantum absorption as its "death".

It is necessary to specify the boundary conditions at the top and bottom of the atmosphere. As it has been mentioned above, solar radiation characterizing with values F_0, ϑ_0, φ_0 incomes to the top. Usually it is assumed $\varphi_0 = 0$ i.e. all azimuths are counted off the solar azimuth and specified $\mu_0 = \cos\vartheta_0$.

As it has been mentioned above solar radiation in the Earth atmosphere consists of direct and scattered radiation. It is accepted not to include the direct radiation to the transfer equation and to write the equation only for the scattered one. The calculation of the direct radiation is elementary accomplished using Beer's Law (1.23). Therefore, present the radiance as a sum of direct and scattered radiance $I(\tau,\mu,\varphi) = I'(\tau,\mu,\varphi) + I''(\tau,\mu,\varphi)$. From the expression for the direct radiance of the parallel beam (1.10) the following is correct $I'(0,\mu,\varphi) = F_0\delta(\mu - \mu_0)\delta(\varphi - 0)$, and it leads to $I'(0,\mu,\varphi) = F_0\delta(\mu - \mu_0)\delta(\varphi)\exp(-\tau/\mu_0)$ for Beer's Law. Substitute the above sum to Eq. 1.24, with introducing the dependence upon value μ_0 and omitting primes $I''(\tau,\mu,\mu_0,\varphi)$, we are obtaining *the transfer equation for diffuse radiation.*

$$\mu\frac{I(\tau,\mu,\mu_0,\varphi)}{d\tau}=-I(\tau,\mu,\mu_0,\varphi)+\frac{\omega_0(\tau)}{4\pi}\int_0^{2\pi}d\varphi'\int_{-1}^{1}x(\tau,\gamma)I(\tau,\mu',\mu_0,\varphi')d\mu'$$
$$+\frac{\omega_0(\tau)}{4\pi}F_0x(\tau,\gamma_0)e^{-\tau/\mu_0} \tag{1.25}$$

Point out that Eq. 1.25 is written only for *the diffuse radiation*. The boundary conditions are taking into account by the third term in the right part of Eq. 1.25. The sense of this term is the yield of the *first order of scattering* to the radiance and the integral term describes the contribution of *multiple scattering*.

The ground surface at the bottom of the atmosphere is usually called *the underlying surface* or *the surface*. Solar radiation interacts with the surface reflecting from it. Hence, the laws of the reflection as a boundary condition at the bottom of the atmosphere should be taken into account. However, it is done otherwise in the radiative transfer theory. As it will be shown in the following section, there are comparatively simple methods of calculating the reflection by the surface if it is obtained the solution of the transfer equation for the atmosphere without the interaction between radiation and surface. Thus, neither direct nor reflected radiation is included to Eq. 1.25. As there is no diffused radiation at the atmospheric top and bottom, the boundary conditions are as follows:

$$I(0,\mu,\mu_0,\varphi)=0\quad \mu>0,\quad I(\tau_0,\mu,\mu_0,\varphi)=0\quad \mu<0.$$

Transfer Eq. 1.25 together with boundary conditions defines the problem of the solar diffused radiance in the plane parallel atmosphere. Nowadays different methods both analytical and numerical are elaborated. Some of them will be considered in the following sections.

1.4 Transformation of the Radiation Transfer Equation

Return to the transfer Eq. 1.25 and transform it. Introduce the following noting in the form:*the average intensity* multiplied to 4π:

$$I(\tau,\mu_0)=\int_0^{2\pi}d\varphi\int_{-1}^{1}I(\tau,\mu,\mu_0,\varphi)d\mu \tag{1.26}$$

the diffuse irradiance:

$$H(\tau,\mu_0)=\int_0^{2\pi}d\varphi\int_{-1}^{1}I(\tau,\mu,\mu_0,\varphi)\mu d\mu \tag{1.27}$$

the K-integral:

$$K(\tau,\mu_0) = \int_0^{2\pi} d\varphi \int_{-1}^{1} I(\tau,\mu,\mu_0,\varphi)\mu^2 d\mu \tag{1.28}$$

The value $[I(\tau,\mu_0) - K(\tau,\mu_0)]$ defines the diffuse radiative flux to the horizontal direction. After integrating the Eq. 1.25 over the viewing angle cosine μ and with taking into account above introduced definitions and Eq. 1.10 – the phase function normalizing, and the definition of the mean cosine of the scattering angle (1.16) it is possible to obtain the system of differential equations:

$$\begin{aligned} \frac{dH(\tau,\mu_0)}{d\tau} &= -[1-\omega(\tau)]I(\tau,\mu_0) + \omega(\tau)F_0\exp(-\frac{\tau}{\mu_0}) \\ 3\frac{dK(\tau,\mu_0)}{d\tau} &= -[3-\omega(\tau)x_1(\tau)]H(\tau,\mu_0) + \omega(\tau)F_0\exp(-\frac{\tau}{\mu_0}) \end{aligned} \tag{1.29}$$

This system will be used below for the derivation of asymptotic and Eddington approaches.

Chapter 2
Special Features of Self-surface (Heat) Radiation Forming

Abstract Radiation of the hypothetical black body is considered. General notations and basic equations are presented for calculation spectral radiation of black and real bodies.

2.1 The Black Body Radiation

It is known from experiments that all matters emit constantly electromagnetic waves. The electromagnetic radiation embraces practically all ranges of wavelength. By its nature this radiation is called self heat radiation, because it arises while molecules transmit at exciting level with kinetic interaction (collisions) with consequently returning to unexciting level with quantum emitting. Thus it is understandable that the intensity of the self heat radiation is to be linked with inner energy of matter that is directly proportional to the temperature and is to depend on physical structure of matter.

Dependencies of forming the self heat radiation field allow obtaining an analytical link between quantities of energy emitted by an object at different wavelength in different directions and object's parameters. But these dependencies are simple only for ideal absorber and emitter of electromagnetic waves, which is blackbody (BB) or perfect radiator.

The blackbody is an hypothetical body that emits the maximal radiation for the temperature, does not reflect or transport the incident energy and absorbs all incident energy falling at all wavelengths and from all directions. The notation of blackbody is the key one for description of heat radiation transfer. The perfect radiator blackbody is used as an etalon for calibration of spectral instruments within spectral IR-ranges.

Max Planck (1858–1947) has assumed two presumptions concerning properties of atom oscillators in 1901 aiming theoretically explain spectral distribution of radiation emitted by heated cavity. Firstly Planck had postulated that the energy of the harmonic oscillator is expressed as $E = nhf$, where f is the oscillator frequency;

I. Melnikova et al., *Remote Sensing of the Environment and Radiation Transfer*,
DOI 10.1007/978-3-642-14899-6_2, © Springer-Verlag Berlin Heidelberg 2012

h is the Planck's constant, and *n* is quantum number, that can be only integer. Later it was shown that in reality the number $n + 1/2$ is right but it does not change Planck's result.

Secondly Planck had supposed that oscillators emit energy not continuously but by portions – quants. These quants are emitted while the oscillator transmits from one quantum condition to another.

These two assumptions allowed Planck to theoretically derive a function that expresses blackbody spectral brightness; it's called now ***Planck's function*** and is very important for description of self atmosphere and surface radiation. The blackbody spectral brightness in the ranges of wave numbers ν and $\nu + d\nu$ is defined by the radiant energy dE, emitted by an blackbody surface element dS during the time interval dt in the solid angle $d\omega$. The blackbody radiation obeys Lambert's law and blackbody surface is ideal diffuse thus the direction of radiation is not important.

2.2 Basic Equations

Then following to Planck's law the blackbody spectral brightness depends only on two variables the absolute temperature T and the wave number ν (or equivalent characteristics λ, f)

$$B_\nu(T) = \frac{a\nu^3}{\exp\left(\frac{b\nu}{T}\right) - 1}, \tag{2.1}$$

where $\nu = \frac{10000}{\lambda}$, cm^{-1} is the wave number (λ is the wavelength, μm);

T is the blackbody absolute temperature, °K;
$a = 1.19105$ W;
$b = 1.43874\ 10^{-2}$ °K/m^{-1}.

In terms of wavelength λ the Planck's function look as follows (Fig. 2.1):

$$B_\lambda(T) = \frac{(a_1\lambda^{-5})}{\exp(\frac{b_1}{\lambda T}) - 1}, \tag{2.2}$$

where $a_1 = 3.74 \cdot 10^{-16}$, W m^2, $b_1 = 1.43874 \cdot 10^{-2}$, °K · m.

In terms of the frequency f the Planck's function is written as

$$B_f(T) = \frac{a_1 c^{-4} f^3}{\exp\left(\frac{b_1 f}{cT}\right) - 1}, \tag{2.3}$$

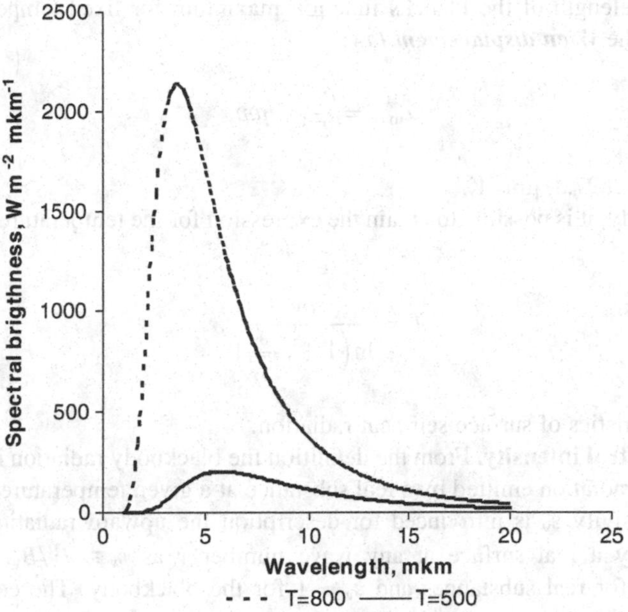

Fig. 2.1 Planck's function for two temperatures 800 and 500 °K

Where $f = c/\lambda$ is the frequency, Hz; ($c = 2.9996 \cdot 10^{14}$ μm/s is the light velocity).

Throughout most of the shortwave range $\lambda < 2$ μm for Earth self heat radiation the exponential term is larger than one, thus ***Wien approximation*** is obeyed by the blackbody spectral brightness:

$$B_\nu(T) = a\nu^3 \exp\left(-\frac{b\,\nu}{T}\right) \tag{2.4}$$

For long waves (λ >100 μm) ***Rayleigh-Jeans approximation*** is true for the blackbody spectral brightness:

$$B_\nu(T) = (a/b)\nu^2 T. \tag{2.5}$$

After integrating the Planck's function over all wave numbers (wavelengths, frequencies) the blackbody *irradiance F(T)* is obtained.

$$F(T) = \sigma T^4 \tag{2.6}$$

The Eq. 2.6 is called ***Stefan-Boltzmann law***, where σ =5.6693 · 10^{-8}, W/(m^2 · °K^4) is the Stefan-Boltzmann constant.

The wavelength of the Plancks function maximum for fixed temperature T is defined by the ***Wien displacement law***:

$$\lambda_{\max} = \frac{c_1}{T}, \quad \mu m, \tag{2.7}$$

where $c_1 = 2897.8$, μm °K.

And finally, it is possible to obtain the expression for the temperature T from the Eq. 2.1:

$$T = \frac{bv}{\ln\left(1 + \frac{av^3}{B_v(T)}\right)}, \tag{2.8}$$

Characteristics of surface self heat radiation.

The spectral intensity. From the definition the blackbody radiation is the upper limit to the radiation emitted by a real substance at a given temperature. The value of the emissivity ε_v is introduced for description the upward radiation intensity $J_v^{\uparrow}$emitted by a real surface at any wave number v as $\varepsilon_v \doteq J_v^{\uparrow}/B_v$. It is clear that $\varepsilon_v < 1$ for real substances and $\varepsilon_v = 1$ for the blackbody. The equation $\varepsilon \equiv J^{\uparrow}/F(T) = J^{\uparrow}/\sigma T^4$ expresses ***gray body*** emissivity.

Then the spectral intensity of the self heat radiation of the surface with the temperature T_s is defined by the following expression:

$$J_v^{\uparrow} = \varepsilon_v B_v(T_s), \tag{2.9}$$

where T_s is the surface temperature; B_v is Planck's function; ε_v is the surface emissivity.

2.3 The Brightness Temperature

The Planck's function allows the numerical describing and conventionally illustrating the spectral distribution of the electromagnetic radiation intensity that is formed by surface or complicated system *atmosphere-surface*. It is reached by assuming that the radiation at any given wave number is formed by the black body at a certain temperature and not by a real substance with a real temperature. Such assumption provides the possibility for every value of intensity J_v to uniquely relate to a certain value of the temperature. This temperature is not a thermodynamic value but only a convenient characteristic for one-to-one describing the spectral distribution of the radiation emitted by the system atmosphere-surface, and it is called ***brightness temperature***. This characteristic is called ***radio-brightness temperature*** at the radio wavelength ranges and ***Rayleigh-Jeans approximation*** (the Eq. 2.5) is used for calculation. The transition from spectral intensity (or brightness) of real substances to their brightness temperature allows the clear plotting of the spectrum in wide ranges and easy correlation of intensity values at different spectral diapasons. The reason is the weak spectral variability of the brightness temperature comparing with intensity. This variability disappeared in the blackbody limit case. For example the blackbody at the temperature of 7,000 K, then brightness temperature $T_\nu = 7{,}000$ K coincides with the thermodynamic and it is constant at all wavelength; the spectral brightness varies over nine orders of magnitude.

The brightness temperature of the heat radiation intensity (including the surface heat radiation) is derived from the relation below

$$B_\nu(T_\nu) \equiv J_\nu^\uparrow \cdot \tag{2.10}$$

The following formula for calculating the brightness temperature T_ν of the self heat radiation intensity of the surface is derived by taking into account Eqs. 2.8, 2.9 and 2.10 :

$$T_\nu = \frac{b\nu}{\ln\left[1 + \frac{a\nu^3}{J_\nu^\uparrow}\right]}. \tag{2.11}$$

By substituting the Eq. 2.9 it is obtained for T_ν:

$$T_\nu = \frac{b\nu}{\ln\left[1 + \frac{a\nu^3}{\varepsilon_\nu B_\nu(T_s)}\right]}. \tag{2.12}$$

It is clear that for $\varepsilon_\nu = 1$ the equality $T_\nu \equiv T_s$ is valid, hence the blackbody brightness temperature does not depend on wave number and equal to thermodynamic temperature of the blackbody surface.

From Eqs. 2.9 and 2.1 the expression for calculating the surface temperature T_s (real thermodynamic value) from measured heat intensity $J_\nu^\uparrow$ is obtained. The assumption of atmosphere absence is taking (for example the temperature of the Moon surface):

$$T_s = \frac{b\nu}{\ln\left[1 + \varepsilon_\nu \frac{a\nu^3}{J_\nu^\uparrow}\right]}. \tag{2.13}$$

The Eq. 2.13 provides the result of remote retrieval of the surface temperature T_s from measuring the surface self heat radiation intensity $J_\nu^\uparrow$. It is necessary a priori knowing the surface emissivity ε_ν and absence of gaseous substance between the surface and instrument. Comparison of Eqs. 2.12 and 2.13 allows understanding the difference between brightness and thermodynamic temperatures.

The sensitivity function. Let us introduce the function S_ν for numerical estimation of the sensitivity of the Planck's function to the temperature variability:

$$S_\nu = \frac{\partial B_\nu[T]}{\partial T}. \tag{2.14}$$

The differentiating of the Eq. 2.1 over the temperature gives the expression for the sensitivity function:

$$S_\nu = \frac{ab\nu^4}{\left[\exp\left(\frac{b\nu}{T}\right) - 1\right]^2}\left[\exp\left(\frac{b\nu}{T}\right)\right]\frac{1}{T}. \tag{2.15}$$

2.4 Practice 1

2.4.1 Objectives

1. Studying the blackbody spectral distribution at different temperatures.
2. Studying the spectral distribution of the Planck's function derivative at different blackbody temperatures.
3. Studying the spectral distribution of the safe heat radiation intensity of real body at different temperatures with numerical simulating spectral variations of its emissivity.
4. Studying the spectral distribution of the brightness temperature of the surface with Planck's function derivative at different blackbody temperatures numerical simulating spectral variations of the surface emissivity.

2.4.2 Software and Set of Input Parameters

1. Computer programs "F_PLANCK", "I_RADT", setup at the directory *dz-*2006*Lab*1.
2. The set of input parameters for programs (Table 2.1).

Table 2.1 Variants of parameter values

Number of the variant	λ_{min}, μm	λ_{max}, μm	T_1, °K	T_2, °K	T_3, °K
1	1	20	550	770	840
2	1	30	440	560	610
3	15	40	260	270	320
4	10	60	200	300	350
5	20	50	150	250	350
6	2	80	100	200	300
7	15	50	270	290	310
8	10	50	300	350	600
9	20	100	230	260	340
10	20	80	300	400	500

2.4.3 *Test Questions*

1. What is definition of the blackbody?
2. What is spectral distribution of the blackbody brightness?
3. What defines the place of spectral brightness maximum?
4. What is the formula for defining the blackbody spectral brightness dependence on wave number and temperature?
5. Does the value of blackbody brightness temperature change on wavelength?
6. Do maximums of spectral brightness and Planck's function derivative over temperature coincide?
7. What parameters does the intensity of the surface heat radiation depend on?
8. Might the surface brightness temperature be equal to its thermodynamic temperature?
9. Derive the formula for the Planck's function derivative over temperature.
10. Demonstrate the validity of the Eq. 2.6 using differentiation of the Eq. 2.3 over wavelength.

2.4.4 *Sequential Steps of the Exercise Implementation*

1. To study the theory with using additional books, pointed in reference list.
2. To take the three variants of input parameters from the Table 2.1 for computer programs "F_PLANCK.exe", "I_RADT.exe". You can use another set of input parameters. Take in mind that the program can operate with wavelength $\lambda \geq 1.0$ μm.
3. To analyze

   – the Planck's function and its derivative over the temperature with using computer program "F_PLANCK.exe" in chosen spectral ranges;
   – the variation of spectral dependence of mentioned functions on temperature;

   Results are demonstrated on the screen and output in file "*fplanck*".
4. To create plots of Planck's functions and its derivative of the temperature (with using Excel)
5. To plot the modeling presentation of the emissivity versus wavelength $\varepsilon(\lambda)$ in Excel in chosen wavelength ranges. The emissivity varies within ranges 0.50–0.97. It is necessary to create the table containing not less 25 values, to plot the emissivity, to approximate the curve with using the polynomial 3rd order trend line, output the corresponded equation at the plot and to fix values of polynomial coefficients for using them in the computer program "I_RADT.exe". (Call attention to the inverse order of the polynomial and input in program "I_RADT.exe" coefficients.).

6. To examine the spectral dependence of surface self heat radiation with using values of the emissivity approximation coefficients obtained in item five. Call special attention to spectral dependence of the surface brightness temperature
7. To plot calculated spectral values, for putting into the final report.

2.4.5 *Requirements to the Report*

Compile the final concise report with elements of theory, resulting pictures, and conclusions that reflects main stages of the work.

Chapter 3
The Direct Calculation of the Absorption Coefficient of Atmosphere Gases with Using Parameters of Absorption Bands Fine Structure

Abstract The selective gaseous absorption is considered. Spectral line broadening is explained. The practice for calculation of the absorption coefficient is described.

Registering the radiation spectral transmission demonstrates a strong selectivity of the absorption in the atmosphere: there are spectral ranges where the radiation is completely absorbed and other ranges where the radiation is nearly not varied. Ranges with strongly absorbed radiation are called *absorption bands*. They are constantly arranged in an appropriate pattern that points to gaseous absorption. The absorption mechanism is different in different spectral ranges and is provoked by quantum nature of the matter. Thus the radiation is considered as a flux of photons with energy $h\nu$.

3.1 The Analysis of Physical Processes of the Interaction

Molecules and atoms of the matter (here atmospheric gases) absorb incoming electro-magnetic radiation and become excited with transition to a higher energetic level. The emission of photons (quanta) leads to the molecule or atom transition to a lower energetic level. It produces a change of energetic level of an electron and rotation and oscillation (vibration) of a molecule. Molecules, atoms and electrons are of energy only corresponding to permitted levels. Hence the process of absorption (emission) is quantum and only lines appropriate to permitted energetic transitions are observed. The amount of energy associated with a photon of radiation is given by

$$E = h\nu = h/\lambda$$

I. Melnikova et al., *Remote Sensing of the Environment and Radiation Transfer*,
DOI 10.1007/978-3-642-14899-6_3, © Springer-Verlag Berlin Heidelberg 2012

where v is the wave number and h is Planck's constant, which is equal to 6.626×10^{-34}Js. It is clear that the decreasing of wavelength λ increases energy of an electro-magnetic wave.

The inner energy of a molecule adds up as follows:

$$E = E_{transl} + E_{electron} + E_{vibrat} + E_{rotat} + E_{el-vibr} + E_{el-rot} + E_{vibr-rot}$$

Continue spectrum is provided only by translational motion (heat) of molecules. Other interaction types are characterized discrete spectrum. Hence, change of molecule energy happens with quantum jump hv. The solution of Schrödinger equation for the wave function (eigenvectors and eigenfunctions) presents the set of discrete values of molecule energy.

The Pauli-Fermi principle determines permitted levels of molecule energy E_i, and the portion $\Delta E_{ij} = E_i - E_j$, of energy changes while molecule transits from the initial level E_i to the level E_j. The Planck's formula $\Delta E_{ij} = hv_{ij}$ determines the molecule absorption (emission) spectrum. Hence, only permitted energy values correspond to only certain wave numbers (wavelengths), forming spectral lines.

Some information concerning the energy of interaction between radiation and mater in different spectral intervals is presented in the Table 3.1. The most energetic quanta are gamma rays which provoke variations of nuclear configuration. Roentgen and UV radiation provides electron transitions from one to other levels. Visible and IR radiation changes vibrational and rotational energy of molecules.

The absorption called by vibrational and rotational and vibrational-rotational transitions is most significant in spectral ranges of the Earth outgoing radiation (maximum at 12 μm).

Ozone bands in UV ranges, the oxygen band 0.76 μm and water vapor bands in visible interval are caused by electron transitions.

The scheme of molecule energetic levels and corresponding transitions is shown in the Fig. 3.1.

Every transition forms an absorption line (emission). Between different levels might be a lot of transition but not all of them are permitted with the Pauli-Fermi principle. The totality of absorption lines provoked by transition between two specific electron levels and different vibrational and rotational levels form an

Table 3.1 Energy types and energy of interaction in different spectral intervals

Radiation	Matter changes	λ, μm	v, cm^{-1}	f, Hz	E, J mol^{-1}
Gamma rays	Change of nuclei configuration	10^{-4}	10^8	3 10^{18}	10^9
Roentgen rays	Electron transition	10^{-2}	10^6	3 10^{16}	10^7
UV and visible	between shells	1	10^4	3 10^{14}	10^5
Infrared	Molecule vibrations	10^2	10^2	3 10^{12}	10^3
Microwave	Molecule rotations	10^4 (1 cm)	1	3 10^{10}	10
Centimeter interval	Changes of electrons	10^6 (100 cm)	10^{-2}	3 10^8	10^{-1}
Meter interval	and nuclear spins	10^8 (10 m)	10^{-4}	3 10^6	10^{-3}

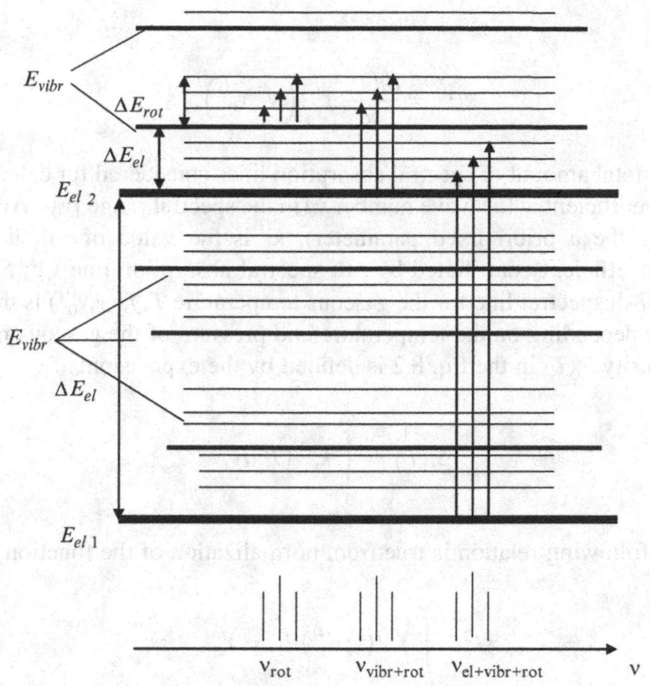

Fig. 3.1 Energy levels of the molecule

electronic absorption band. There are vibrational or rotational absorption bands when the electronic energy is not changed. Spectra are called also electronic, vibrational or rotational. The complicated character of molecule total energy variations is responsible for complicated structure of molecule spectrum.

The spectral line broadening is observed as it will be considered below. The contour of spectral line $f^{(i)}\left(\nu, \nu_0^{(i)}\right)$ describes this broadening.

3.2 Infrared Spectral Range

Only carbon dioxide is considered here as atmospheric gaseous component because of invariable content. The expression for calculating the monochromatic absorption coefficient k_ν might be written as the sum over all lines within the spectral ranges $\Delta\nu$:

$$k_\nu = \sum_{i=1}^{N} k_\nu{}^{(i)}, \tag{3.1}$$

where

$$k_v{}^{(i)} = \frac{S_i(T)}{\pi} f^{(i)}\left(v,\ v_0^{(i)}\right). \tag{3.2}$$

and N is the total amount of spectral absorption lines considered for calculating the absorption coefficient at the wave number ν (in the spectral range $[v_0 - \Delta v,\ v_0 + \Delta v\ ]$, where Δv is the a priori fixed parameter); $k_v^{(i)}$ is the value of i-th item of the absorption coefficient contributed by i-th spectral absorption line $v_0^{(i)}$; $S_i(T)$ is the intensity of i-th spectral line for the gaseous temperature T; $f_i\ (v, v_0^{(i)})$ is the spectral line contour depending on the temperature and pressure of the gaseous medium.

The intensity $S_i(T)$ in the Eq. 8.2 is defined by the expression:

$$S_i(T) = \int_0^{\infty} k_v^{(i)}\ (T)dv, \tag{3.3}$$

And the following relation is true from normalization of the function $f_i\ (v,\ v_0^{(i)})$:

$$\int_0^{\infty} f^{(i)}(v, v_0^{(i)})dv = 1 \tag{3.4}$$

The dependence of the intensity S_i on temperature might be expressed with the following relation:

$$S_i(T) = S_i(T_0)\frac{Q_\vartheta(T_0)}{Q_\vartheta(T)}\left(\frac{T_0}{T}\right)^j \exp\left[-\frac{1439\, E''_i\, (T - T_0)}{T\, T_0}\right]. \tag{3.5}$$

where $S_i(T_0)$ is the i-th spectral line intensity for the temperature T_0; E'' is the energy ([cm^{-1}]) of the lowest level, from which the transition happens; $Q_\vartheta(T)$ is so called *the vibrational statistical sum*; j is the parameter depending on molecule type. Values of parameters j и $Q_\vartheta(T)$ including in the Eq. 3.5 are presented in the Table 3.2 for defining the line contour $f_i(v, v_0^{(i)})$ as a function of parameters determined by gaseous

Table 3.2 Values of coefficients j and Q_ϑ for the set of molecules

Molecule	j	$Q_\vartheta(T)$			
		200 K	250 K	266 K	325 K
H_2O	1.5	1.000	1.000	1.000	1.001
CO_2	1.0	1.0192	1.0502	1.0931	1.1269
O_3	1.5	1.007	1.022	1.046	1.066
N_2O	1.0	1.030	1.072	1.127	1.170
CO	1.0	1.000	1.000	1.000	1.000
CH_4	1.5	1.000	1:002	1.007	1.011

media physical state while applying Eqs. 8.1, 8.2 and 8.5 for calculating the absorption coefficient k_ν.

The absorption (emission) line is not a just line. Certain mechanisms govern the spectral line broadening. The most significant are broadening by collisions (pressure) and by Doppler's effect.

Line broadening by pressure. *Lorenz's contour* is used for quantitative description of the shape of *i*-the absorption line to account for this broadening:

$$f_L^{(i)}\left(\nu,\ \nu_0^{(i)}\right) = \frac{\alpha_L^{(i)}}{\left(\nu - \nu_0^{(i)}\right)^2 + \left(\alpha_L^{(i)}\right)^2}, \tag{3.6}$$

where $\alpha_L^{(i)}$ is *the* halfwidth of absorption line at the wave number $\nu_0^{(i)}$ (the width of the contour at the half maximum level). The *i*-th spectral line halfwidth $\alpha_L^{(i)}$ is the function of the atmospheric pressure and temperature: $\alpha_L^{(i)} = \alpha_L^{(i)}(p,T)$. Often the expression for this function is assumed:

$$\alpha_L^{(i)}(p,T) = \alpha_L^{(i)}(p_0,T_0)\frac{p}{p_0}\left(\frac{T_0}{T}\right)^{\frac{1}{2}}, \tag{3.7}$$

where $\alpha_L^{(i)}(p_0,T_0)$ is the *i*-th spectral line halfwidth at standard values of the pressure p_0 and temperature T_0.

Doppler's broadening. The broadening by pressure is possible to neglect when the pressure is low (the rarefaction gas). But molecules are of high velocity and if the molecule emits at the wave number ν_0, and the component of the velocity in the viewing direction is ϑ. Then from the observer point of view the molecule emits at the wave number ν that defines as:

$$\nu = \nu_0\left(1 \pm \frac{\vartheta}{c}\right). \tag{3.8}$$

Velocities of molecules ϑ obey to Maxwell-Boltzmann's distribution, what provokes various shifts of the wave numbers ν, i.e. spectral line broadening. The corresponding absorption coefficient is calculated with the following formula:

$$k_\nu^{(i)} = \frac{S_i(T)}{\alpha_D^{(i)}\sqrt{\pi}}\exp\left[-\frac{\left(\nu - \nu_0^{(i)}\right)^2}{\alpha_D^{(i)}}\right] = S_i(T)\,f_D^{(i)}\left(\nu,\ \nu_0^{(i)}\right). \tag{3.9}$$

The parameter $\alpha_D^{(i)}$, included to the Eq. 3.9 is called *Doppler's width of the spectral line*:

$$\alpha_D^{(i)}(T) = \frac{\nu_0^{(i)}}{c}\left(\frac{2\,kT}{m}\right)^{\frac{1}{2}}, \tag{3.10}$$

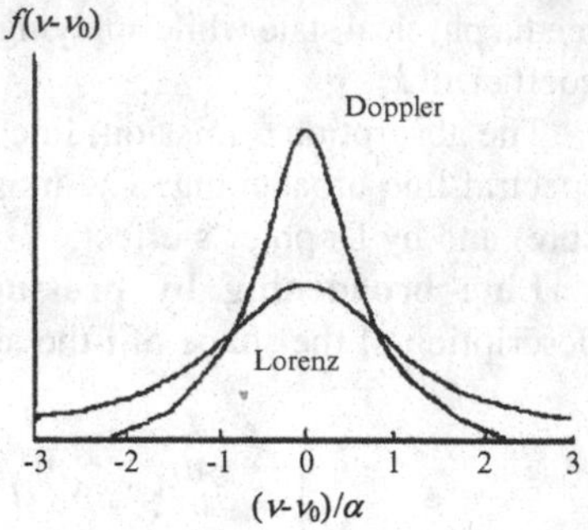

Fig. 3.2 Comparison of Lorenz's and Doppler's contours *1* – Lorenz's contour, *2* – Doppler's contour

where m is the molecule mass; k is Boltzmann's constant; T is the absolute temperature. Doppler's halfwidth of the absorption line is $\alpha_D^{(i)}\sqrt{\ln 2}$.

The comparison of Lorenz and Doppler's contours is shown in the Fig. 3.2.

Voigt's contour. The mutual accounting for both Lorenz and Doppler's broadening in calculating the function $f_i(\nu, \nu_0^{(i)})$ leads to Lorenz-Doppler's contour that is called the Voigt's contour. The absorption coefficient in the Eq. 8.1 is assumed as

$$k_\nu^{(i)} = k_{0\nu}^{(i)} f_i\left(\nu, \nu_0^{(i)}\right), \tag{3.11}$$

where

$$k_{0\nu}^{(i)} = \sqrt{\frac{\ln 2}{\pi}} \frac{S_i(T)}{\alpha_D}. \tag{3.12}$$

The expression for the contour of i-th spectral line is following:

$$f_i\left(\nu, \nu_0^{(i)}\right) = \frac{a}{\pi} \int_{-\infty}^{\infty} \frac{e^{-x^2}}{a^2 + (w - x)^2} dx, \tag{3.13}$$

where

$$a = \sqrt{\ln 2} \frac{\alpha_L^{(i)}}{\alpha_D^{(i)}}, \tag{3.14}$$

$$w = \sqrt{\ln 2} \frac{\left(\nu - \nu_0^{(i)}\right)}{\alpha_D^{(i)}}. \tag{3.15}$$

3.3 The Microwave Spectral Interval

Significant absorption (lines) bands in microwave interval are possessed by water vapor H_2O and oxygen molecule O_2. The single slight water vapor broadening line 22.235 GHz dominates at the frequency lower than 40 GHz. The atmosphere is rather transparent at the frequency near 31.4 GHz. This interval is the window between resonance line of water vapor and strongly absorbing oxygen band centered at 60 GHz. The single absorption line of the oxygen molecule is at the frequency 118.75 GHz that determines absorption of microwave radiation in this spectral range. The strong line near 183 GHz provides prevailing H_2O absorption at frequencies more than 120 GHz. The Fig. 3.3 illustrates the transmission dependency on frequency in the microwave interval.

The absorption of electro-magnetic radiation is defined in the frequency interval 110–120 GHz by the following components:

- Oxygen molecules band formed by resonance and non-resonance lines in the wavelength range near 5 mm;
- Water vapor absorption band;
- Water droplets absorption (by cloud, rain).

The absorption coefficient of clear atmosphere without water droplet absorption might be calculated by the following expression:

$$G(v_i, h_j) = G_{O_2}(v_i, h_j) + G_{H_2O}(v_i, h_j), \tag{3.16}$$

where h_j is the altitude ($j = 1.2,\ldots, N$), v_i is the frequency ($i = 1.2,\ldots, M$).

Note that it is the frequency (GHz) that is noted here ν and not the wave number (cm^{-1}) as above!

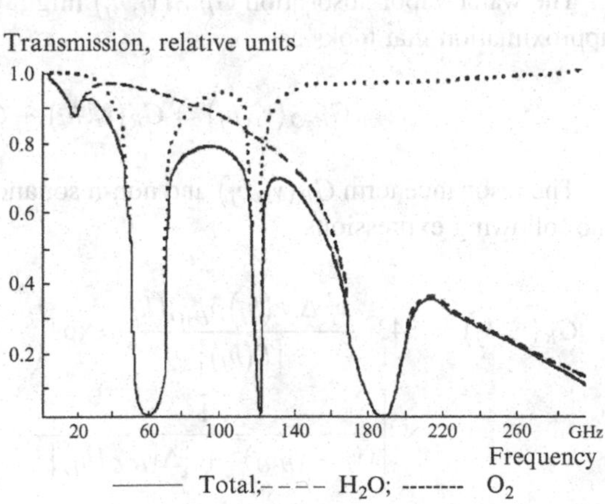

Fig. 3.3 Dependence of transmission of the atmosphere at the frequency in microwave interval

The formula accounting for resonance absorption in the oxygen spectral line at the wavelength $\lambda = 2.53$ mm and the contribution of oxygen molecule absorption lines near $\lambda = 5$ mm for calculating the first component G_{o_2} of absorption coefficient:

$$G_{O_2}(\nu_i, h_j) = 1.2305 \frac{\nu_i^2 p(h_j)}{\left[T(h_j)\right]^3} \exp\left(-\frac{4.14}{T(h_j)}\right) \times \frac{4\nu_i^2 \Delta\nu(h_j)}{\left(\nu_{O_2}{}^2 - \nu_i{}^2\right)^2 + 4\nu_{O_2}{}^2\left[\Delta\nu_{O_2}(h_j)\right]^2}. \tag{3.17}$$

The central frequency of the line is $\nu_O = 118.750343$ GHz; $p(h_j)$ is the atmospheric pressure (mmHg) at the altitude h_j; $T(h_j)$ is the temperature (K) at the altitude h_j; The dimensions of the value G_{O_2} is [1/km].

The halfwidth value of oxygen absorption line at the wavelength $\lambda = 2.53$ mm and at the altitude h_j is defined as:

$$\Delta\nu_{O_2}(h_j) = \left\{\left[\Delta\nu_p(h_j)\right]^2 + \left[\Delta\nu_D(h_j)\right]^2\right\}^{\frac{1}{2}}. \tag{3.18}$$

Where $\Delta\nu_p$ and $\Delta\nu_D$ are broadening by pressure and Doppler broadening:

$$\Delta\nu_D(h_j) = 7.5210^{-6}\sqrt{T(h_j)}; \tag{3.19}$$

$$\Delta\nu_p(h_j) = \alpha p(h_j)[0.21 + 0.78\,\beta]\left[\frac{300}{T(h_j)}\right]^C. \tag{3.20}$$

The value $\alpha = 2.131*10^{-3}$(GHz/mmHg) in the last equation is the coefficient of broadening line by pressure, and the value $\beta = 0.75$ is the coefficient of collisions effectiveness for molecules N_2 and O_2 comparing with collisions between O_2 and O_2; The parameter $C = 0.9$ is the temperature coefficient.

The water vapor absorption $G_{H_2O}(\nu_i, h_j)$ might be calculated with the empirical approximation that looks as:

$$G_{H_2O}(\nu_i, h_j) = G_R(\nu_i, h_j) + G_N(\nu_i, h_j). \tag{3.21}$$

The resonance term $G_R(\nu_i, h_j)$ and non-resonance term $G_N(\nu_i, h_j)$ are defined by the following expressions:

$$G_R(\nu_i, h_j) = 343 \frac{\nu_i^2\,\Delta\nu(h_j)\rho_{H_2O}(h_j)}{\left[T(h_j)\right]^{\frac{5}{2}}} \exp\left(-\frac{644}{T(h_j)}\right) \times \left[\frac{1}{(\nu_i - \nu_{H_2O})^2 + \left[\Delta\nu_{H_2O}(h_j)\right]^2} + \frac{1}{(\nu_i + \nu_{H_2O})^2 + \left[\Delta\nu_{H_2O}(h_j)\right]^2}\right], \tag{3.22}$$

$$G_N(\nu_i, h_j) = 2.55 \cdot 10^{-3} \frac{\nu_i^2 \Delta\nu_{H_2O}(h_j)\rho_{H_2O}(h_j)}{[T(h_j)]^{\frac{3}{2}}}. \quad (3.23)$$

Where $\nu_{H_2O} = 22.235$ GHz; $\rho_{H_2O}(h_j)$ [g/m^3] is the water vapor density at the altitude h_j; the dimensions of the value G_R and G_N is [1/km].

Calculating the halfwidth of absorption H_2O line at the frequency $\nu = 22.235$ GHz is done according to the following:

$$\Delta\nu_{H_2O}(h_j) = 0.126 \frac{p(h_j)}{[T(h_j)]^{0.626}} \left[1 + 0.011 \frac{\rho_{H_2O}(h_j)T(h_j)}{p(h_j)}\right]. \quad (3.24)$$

If there is no observational information then simulating the vertical profile of absolute humidity looks as:

$$\rho_{H_2O}(h_j) = 7.5 \exp\left(-\frac{h_j}{1.74}\right). \quad (3.25)$$

If the specific humidity profile q (g/g) is known the value ρ_{H_2O}(g/m^3) is calculated according to:

$$\rho_{H_2O}(h_j) = \frac{288.87 p(h_j) q(h_j)}{T(h_j)\,[0.62197 + 0.37803 q(h_j)]}. \quad (3.26)$$

3.4 Practice 2

3.4.1 Objectives

1. To study the approach for direct calculation of the absorption coefficient of atmospheric gaseous components on the base of using data of the fine structure parameters and getting skills of forming data and accomplishing computer calculations.
2. To study the fine structure parameters dependence on the atmospheric pressure and temperature within the spectral IR interval.
3. To study the approach for calculating characteristics of the atmosphere absorption in case studies of values of the absorption coefficient within oxygen absorption at the wavelength $\lambda = 2.53$ mm.
4. To study the following characteristics in IR and Microwave intervals basing on the analysis of the results of calculating the atmosphere absorption coefficient:

   – the absorption coefficient dependence on the wavelength;
   – the absorption coefficient dependence on meteorological parameters.

3.4.2 Software and Set of Input Parameters

1. The table of the fine structure parameters of atmospheric gases in fixed interval of wave numbers. The ready file might be used in spite of the table.
2. Data of the aerological temperature-wind sounding preparing with the computer program "V_AERDAT.exe".
3. Computer programs "V_CONTUR.exe", "CONTUR_1.exe", "CONTUR_2.exe" and "CONTUR_3.exe" "SVCRAD.exe" written in the computer language Basic, are situated in the directory *lab11*. Results are provided in a file "*contour.dat*".
4. Plotting results with EXCEL editor.

3.4.3 Test Questions

1. What characteristics of the physical state of atmosphere govern the value of absorption coefficient in spectral ranges near 2.5 mm?
2. What altitude ranges does the maximum of the absorption coefficient in the considered spectral interval situated?

3.4.4 Sequential Steps of the Exercise Implementation

1. To study theoretical base of forming absorption lines of atmospheric gaseous components.
2. Get the necessary data on the characteristics of the fine structure of absorption line in the Infrared wavelength range from the teacher. If the data is contained in the hard drive, then the student is given the file name and its location. If the data is contained in the summary table of the fine structure of the various gaseous components of the atmosphere, the its necessary to decode the data from this table and enter them into the PC (written in the data file). The summary table contains 12 columns, which is explained by the structure of the Table 3.3.
   The parameter a_i containing in coded form the information about absorption line halfwidth $\alpha_L^{(i)}$, molecular weight μ and gaseous component code N are in columns 4, 8 and 12 of the Table 3.4:
   $a_i = \alpha_L^{(i)} \times 10^7 + \mu \times 10^2 + N.$
   Decoding of gaseous component information is presented in the Table 3.4.

Table 3.3 The structure of presenting the fine structure parameters of absorption lines in a summary table

$\nu_0^{(i)}$	$S_i(T_O)$	E''_i	a_i	$\nu_0^{(i)}$	$S_i(T_O)$	E''_i	a_i	$\nu_0^{(i)}$	$S_i(T_O)$	E''_i	a_i
1	2	3	4	5	6	7	8	9	10	11	12

Table 3.4 Codes of gaseous components

Codes	Gaseous component
1	H_2O
2	CO_2
3	O_3
4	N_2O
5	CO
6	CH_4
7	O_2

3. To prepare tables of the fine structure parameters for every component (if there is not in file). To prepare data files with computer program "V_contur" (separate file for every component), containing values of frequencies $\nu_0^{(i)}$, intensities $S^{(i)}$, halfwidths $\alpha^{(i)}$ and energy of lowest energy E^i. Every file has to contain the information of one gaseous component. Data input strictly in ascending order of wave numbers $\nu_0^{(i)}$. File names are to be composed in accordance with instructions below.
4. To plot the absorption line intensity for every component as a straight vertical line. On the abscissa OX values of wave number and on the ordinate OY intensities of absorption lines in logarithmic scale are plotted (the scale of the axes OX is to be similar). Plotting might be done in EXCEL editor.
5. To study the character of the pressure and temperature impact to:

   - the intensity and halfwidth of absorption line;
   - the spectral variations of the absorption coefficient.

   The study is accomplished with the computer program "CONTUR_1.exe", that realizes the calculation of the absorption coefficient of the single absorption line. Only Lorenz's contour is considered.
6. To calculate the spectral dependence of every gaseous component absorption coefficient with the program "CONTUR_2.exe", the file of aerological sounding data and sequentially pointing file names of every component.
   Calculation is accomplished separately for 3–4 temperature values and pressure in atmospheric layers

   - 1,000 ÷ 700 GPa,
   - 700 ÷ 300 GPa,
   - 300 ÷ p_{min} GPa.

   Positions of corresponding levels input with pointing numbers in tables of aerological sounding. The table is output to the screen after every running of the program "CONTUR_2.exe". In every case, the resulting plot of the spectral absorption coefficient is written to a file "*contour2.dat*". All obtained results are to be plotted and included to the final report.
7. To calculate the vertical profile of the absorption coefficient of every component with the program "Contur_3" and file of aerological sounding data for 3 wave numbers (for strong, moderate and weak absorption).

8. To calculate profiles of total absorption coefficient and its constituents for one fixed frequency in ranges 110–120 GHz with the program "Svcrad" and file of aerological sounding data.
9. To calculate the spectral dependence of the clear absorption coefficient in frequency ranges 110–120 GHz with step 0.5 GHz with the program "Svcrad" for fixed altitude level.

3.4.5 *Requirements to the Report*

To prepare the brief report including elements of the theory, all results, and conclusions.

3.4.6 *Additional Formulas and Relations*

1. G [1/km] = [ln (10)/10] × G [dB/km] = 0.230259 × G [dB/km].
2. The absorption coefficient within microwave interval might be calculated with the Van-Fleck-Veiskon formula

$$G_{O_2}(\nu_i, h_j)[1/km] = 1.2305 \frac{\nu_i^2 p(h_j)}{[T(h_j)]^3} \exp\left(-4.14 \frac{1}{T(h_j)}\right) \times \left[\frac{\Delta\nu(h_j)}{(\nu_i - \nu_{O_2})^2 + [\Delta\nu(h_j)]^2} + \frac{\Delta\nu(h_j)}{(\nu_i + \nu_{O_2})^2 + [\Delta\nu(h_j)]^2}\right]$$

3. The following formula for calculation non resonance absorption by water vapor in the spectral interval $\lambda = 2.2$–3.0 mm:

$$G_{H_2O}[dB/km] = 1.73 \cdot 10^{-3} \rho_{H_2O}(h_j) \left[\frac{T(h_j)}{300}\right]^{\frac{5}{2}} \frac{p(h_j)}{760} \left(\frac{1}{\lambda}\right)^{2.4},$$

where ρ_{H_2O} is the absolute humidity [g/m^3];
λ is the wavelength [cm];
$p(h_j)$ is the pressure at the altitude $P_\nu(u) = \frac{1}{\Delta\nu} \int\limits_{\Delta\nu} e^{-k_\nu u}\, d\nu$, [mmHg]

Chapter 4
Calculating Transmission Functions with Modeling Absorption Bands of Atmospheric Gases

Abstract Molecular absorption bands are modeled. Elsasser's regular and statistic (Goody's) models are considered. The algorithm for calculation transmission function is presented. The Practice for the transmission function is described.

4.1 The Individual Spectral Line

The ideal case of the atmosphere containing only one gas characterized by only one spectral absorption line is considered. The contour of the line, which generally might depend on many factors, is assumed here to be possessing Lorenz's shape. Then the absorption coefficient k_ν at the wave number ν is possible to express as:

$$k_\nu = \frac{S}{\pi} \frac{\alpha}{(\nu - \nu_0)^2 + \alpha^2} = Sf(\nu - \nu_0), \tag{4.1}$$

where S is the intensity of the absorption line (depends on temperature); ν_0 is the frequency defining the spectral line position in spectrum; α is the halfwidth depending on pressure and temperature; The function $f(\nu - \nu_0)$ is the spectral absorption line contour. With reasoning the homogeneous (over altitude) atmosphere and the constant absorption coefficient $k_\nu = const$, the expression for the transmission function in the spectral interval $\Delta\nu$ could be written as follows:

$$P_\nu(u) = \frac{1}{\Delta\nu} \int\limits_{\Delta\nu} e^{-k_\nu u} d\nu\,, \tag{4.2}$$

where u is the integral content of absorbing gas over the radiation path.

Thus, the presentation of the exponent is needed for integrating in the Eq. 4.2.

I. Melnikova et al., *Remote Sensing of the Environment and Radiation Transfer*, DOI 10.1007/978-3-642-14899-6_4, © Springer-Verlag Berlin Heidelberg 2012

Let the viewing zenith angle be θ, and the radiation passes between altitude levels z_1 and z_2. Then the value $u = u(\theta, z_1 z_2)$ is defined for the heterogeneous atmosphere with the altitudinal profile of absorbing component density $\rho(z)$ as:

$$u(\theta,\, z_1, z_2) = \int_{z_1}^{z_2} \rho\,(x)\, \frac{dx}{\cos\,(\theta)}. \tag{4.3}$$

It is possible obtaining the following formula for the absorption coefficient with substitution of the Eq. 4.1 to the Eq. 4.2:

$$P_\nu(u) = \frac{1}{\Delta\nu} \int_{\Delta\nu} \exp\left[-\frac{S}{\pi} \frac{\alpha}{(\nu-\nu_0)^2 + \alpha^2}\, u \right] d\nu. \tag{4.4}$$

Two important limiting cases concerned weak and strong absorption follow from the Eqs. 4.2 and 4.4. It is true $(S\ u/\alpha) \to 0$ for the weak absorption and the absorption function is defined in the fashion:

$$A_\nu = 1 - P_\nu \approx (S/\Delta\nu)u. \tag{4.5}$$

The absorption function A_ν is proportional to the value u, and the ranges of low values $(S\ u/\alpha)$ is called *the linear absorption* region.

The strong absorption corresponds to *the strong line* $(S\ u/\alpha)\to\infty$, thus the value $\alpha \ll 1$ and the absorption function A_ν is derived in the form:

$$A_\nu = 1 - P_\nu \approx \left(2\sqrt{S\alpha/\Delta\nu}\right)\sqrt{u} \tag{4.6}$$

Here the absorption is proportional to the square root of the value u and ranges of high values $(S\ u/\alpha)$ is called region of *the square root law*. The absorption dependence on the absorbing gas content u is shown in the Fig. 4.1.

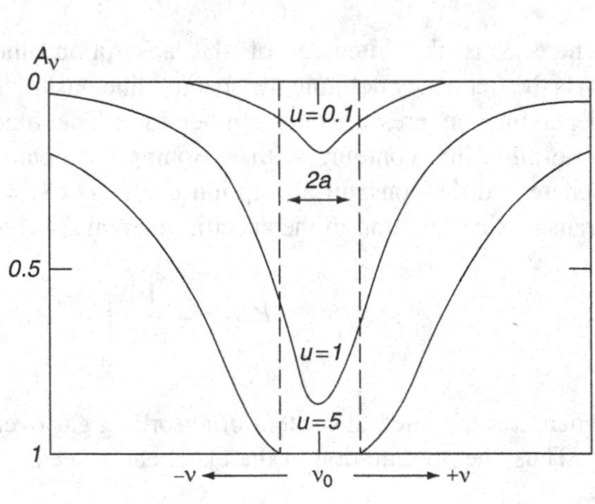

Fig. 4.1 The absorption function A_ν in gaseous medium in spectral line with different values *u*

4.2 Elsasser's Model of the Regular Molecular Band

Elsasser W.M. has in 1938 observed lines evenly distributed in spectral range of CO_2 absorption band 15 μm and proposed this model that consists of periodical lines. The band is sometimes called Elsasser's band.

Let separate absorption line repeat periodically (or regular) as the Fig. 4.2 demonstrates and the corresponded absorption coefficient be defined with the Eq. 4.1. Then the absorption coefficient could be calculated with the following formula, when displacing from any line centre to the value ν by the wave number scales.

$$k_\nu = \sum_{i=-\infty}^{\infty} \frac{S\alpha/\pi}{(\nu - i\delta)^2 + \alpha^2}, \tag{4.7}$$

where δ is the distance between centers of lines.

It can be shown basing on Mittag-Leffler's theorem that an infinite sum might be expressed with trigonometric and hyperbolic functions in the following manner:

$$k_\nu = \frac{S}{\delta} \frac{\sinh \beta}{\cosh \beta - \cos \gamma}, \tag{4.8}$$

where

$$\beta = 2\pi\alpha/\delta \text{ and } \gamma = 2\pi\nu/\delta. \tag{4.9}$$

The transmission function might be expressed after certain transformations as:

$$P_\nu = \int_z^\infty \exp(-z)\mathrm{cth}\beta J_0(iz/\sinh \beta dz), \tag{4.10}$$

where $y = Su/(\delta \sinh \beta)$, $z = y \sinh\beta$, and J_0 is redacted Bessel's function of the first kind and zeroth order. The Eq. 4.10 defines the Elsasser's transmission function.

Certain approximations and simplicities could be done in the considered model, namely, the transmission function is transformed when $\alpha \ll \delta$ и $\beta \to 0$ in the manner:

$$A_\nu = \frac{2}{\sqrt{\pi}} \int_0^x \exp(-x^2)dx = \mathrm{erf}(x) = \mathrm{erf}\left(\frac{\sqrt{\pi S\alpha u}}{\delta}\right), \tag{4.11}$$

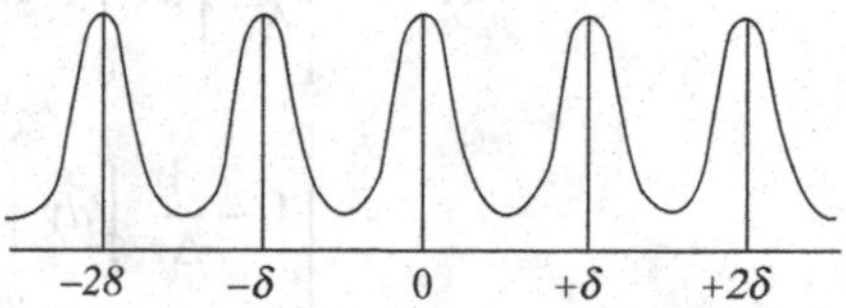

Fig. 4.2 Elsasser's regular model of the absorption band

where

$$x = z\beta/2. \tag{4.12}$$

And the absorption function A_ν becomes rather simple for small values x

$$A_\nu = 2x/\sqrt{\pi} = 2\sqrt{S\alpha u/\delta}. \tag{4.13}$$

The Eq. 4.13 determines the region of the square root low introduced with the Eq. 4.6.

4.3 The Statistic Molecule Band Model (Goody's Model)

Goody R.M. studied in 1952 the water vapor rotational band and has found that the random position of spectral lines is the only feature of the absorption lines spectral distribution in spectral intervals $\Delta\nu$ wider than 25 cm^{-1}. It allows the possibility of analytically calculating the absorption coefficient basing on the random distribution of absorption lines, characterized by known statistical properties (distribution functions).

Let the spectral interval $\Delta\nu$ contain n lines spaced at average distance δ, and it be true $\Delta\nu = n\delta$. Assume the discrete uniform distribution of the line location within the spectral interval $\Delta\nu$. Introduce the function $P(S_i)$ determining the probability of i-th line possessing the intensity S_i, and it is normalized in according with:

$$\int_0^\infty P(S)dS = 1. \tag{4.14}$$

Then the mean function value over the interval $\Delta\nu$ is found with averaging the absorption coefficient over all intensities and all line locations:

$$P_\nu = P_\nu = \frac{1}{(\Delta\nu)} \int_{\Delta\nu} d\nu_1 \ldots \int_{\Delta\nu} d\nu_n \times \int_0^\infty P(S_1)\, e^{-k_1 u} dS_1 \quad \ldots \quad \int_0^\infty P(S_n)\, e^{-k_n u} dS_n, \tag{4.15}$$

where k_n is the absorption coefficient of n-th line. All integrals are similar that leads to:

$$\begin{aligned} P_\nu &= \left[\frac{1}{\Delta\nu} \int d\nu \int_0^\infty P(S)\, e^{-ku} dS \right]^n \\ &= \left[1 - \frac{1}{\Delta\nu} \int d\nu \int_0^\infty P(S)\, (1 - e^{-ku}) dS \right]^n . \end{aligned} \tag{4.16}$$

Because of assuming $\Delta v = n\delta$ and known relation $(1 - x/n)^n \to \exp(-x)$ the Eq. 4.16 can be shown to approach the exponential function for large value n. Thus, it yields:

$$P_v \cong \exp\left\{ -\frac{1}{\Delta v} \int_0^\infty P(S) \left[\int_{\Delta v} \left(1 - e^{-ku}\right) dv \right] dS \right\}. \tag{4.17}$$

It is possible to use different function for describing spectral lines distribution. Consider here the simple function that is the Poisson's distribution:

$$P(S) = \bar{S}^{-1} \exp(S/\bar{S}), \tag{4.18}$$

where $\bar{S}$ is the mean line intensity. After introducing Lorenz's contour for the absorption coefficient k_v to the Eq. 4.17 and integrating over line intensities and wave numbers v from $-\infty$ till $+\infty$ the final result is obtained.

$$P_v = \exp\left[-\frac{\bar{S}u}{\delta} \left(1 + \frac{\bar{S}u}{\pi\alpha}\right)^{-\frac{1}{2}} \right]. \tag{4.19}$$

Note, that the transmission function obtained for random model can be expressed as a function of only two parameters, $\bar{S}/\delta$ and $\pi\alpha/\delta$, for given value u. These two parameters could be found by forcing the experimental or theoretical quantum-mechanical data by the random model for specified line. Take note that calculating simplicity and relatively high accuracy provides considerable current using the random model in problems of remote sensing and the estimation of atmosphere radiation cooling.

Consider the random model for cases of strong and weak absorption. The equivalent width of n lines is to be defined by:

$$W = \frac{1}{n} \sum_{i=1}^{n} W_i = \int_0^\infty P(S) \left[\int \left(1 - e^{-ku}\right) dv \right] dS = \bar{S}u \left(1 + \frac{\bar{S}u}{\pi\alpha}\right)^{-\frac{1}{2}}, \tag{4.20}$$

For the weak absorption $\bar{S}u/\pi\alpha << 1$, it gives the relation:

$$\frac{\bar{S}u}{\delta} = \frac{1}{\Delta v} \sum S_i u, \tag{4.21}$$

where $\Delta v = n\delta$, and S_i is the intensity of the separate i-th line.

For the case of strong absorption $\bar{S}u/\pi\alpha >> 1$, and the result is obtained

$$\frac{\bar{S}}{\delta} = \frac{\sum S_i}{\Delta v}, \quad \frac{\alpha \pi \bar{S}}{\delta^2} = \left(\frac{2 \sum \sqrt{S_i \alpha_i}}{\Delta v} \right)^2. \tag{4.22}$$

Parameters of the statistical model at the temperature $T = 260$ K are presented in the Table 4.1 for the following transmission function versus wave number ν and atmospheric parameters u and p:

$$P_{\Delta\nu_i}(\nu_i, u, p) = \exp\left(\frac{-\beta_i u}{\sqrt{1+\alpha_i u/p}}\right). \tag{4.23}$$

Parameters of the statistical model were obtained basing on the statistical processing experimental data for the 6.3 μm vibrational-rotational water vapor

Table 4.1 Parameters of the statistical model for $T = 260$ K $P_{\Delta\nu_i}(\nu_i, u, p) = \exp\left(\frac{-\beta_i u}{\sqrt{1+\alpha_i u/p}}\right)$

Spectral interval number	ν, cm^{-1}	Δν, cm^{-1}	Gas	α_i, 10^{-3} cm^2 GPa/g	β_i, cm^2/g
1	70	140	H_2O	30 405.1	4 180.5
2	210	140	H_2O	69 058.0	6 278.2
3	380	200	H_2O	12 565.7	672.1
3	380	200	N_2O	52.8	28.5
4	530	100	H_2O	203.5	11.9
4	530	100	N_2O	153.3	206.5
5	610	60	H_2O	129.5	7.7
5	610	60	CO_2	79.7	46.7
6	670	60	H_2O	35.5	1.9
6	670	60	CO_2	2 644.0	1 975.7
7	730	60	H_2O	14.1	0.72
7	730	60	CO_2	60.5	27.2
8	870	220	H_2O	2.6	0.11
9	1,020	80	H_2O	0.17	0.01
9	1,020	80	O_3	252.0	1 938.7
10	1,120	120	H_2O	2.5	0.15
10	1,120	120	O_3	20.8	45.6
10	1,120	120	N_2O	68.8	27.3
11	1,210	60	H_2O	4.9	0.52
11	1,210	60	CH_4	537.1	90.4
11	1,210	60	N_2O	65.1	32.4
12	1,270	60	H_2O	19.1	3.1
12	1,270	60	CH_4	5 378.5	1 325.9
12	1,270	60	N_2O	1 799.2	1 616.7
13	1,330	60	H_2O	136.1	22.8
13	1,330	60	CH_4	9 794.8	2 800.1
13	1,330	60	N_2O	1 227.2	557.6
14	1,390	60	H_2O	895.8	230.5
14	1,390	60	CH_4	685.3	109.4
15	1,550	260	H_2O	7 001.4	1 582.5
15	1,550	260	CH_4	71.4	10.9
16	1,940	520	H_2O	4 486.7	238.4
17	2,230	60	N_2O	14 140.8	10 937.5

band and quantum-mechanical data for the rotational H_2O band, and 15 μm CO_2 band are shown in the Table 4.2. Parameters of the statistical model for the 9.6 μm ozone band are calculated for the spectral interval 1,000–1,060 cm^{-1} at the temperature 233K and $\Delta v = 6.5\ cm^{-1}$ and data are also presented in the Table 4.2. The following expression is used for the transmission function calculation:

$$P_v = \exp\left[-\frac{\bar{S}u}{\delta}\left(1+\frac{\bar{S}u}{\pi\alpha}\right)^{-\frac{1}{2}}\right]. \quad (4.24)$$

Table 4.2 Parameters of the statistical model in the spectral IR interval $P_v = \exp\left[-\frac{\bar{S}u}{\delta}\left(1+\frac{\bar{S}u}{\pi\alpha}\right)^{-\frac{1}{2}}\right]$

Spectral interval, cm^{-1}	$\overline{S}/\delta$, cm^2/g	$\pi\alpha/\delta$
H_2O rotational band		
40–160	7210.30	0.182
160–280	6024.80	0.094
280–380	1614.10	0.081
380–500	139.03	0.080
500–600	21.64	0.068
600–720	2.919	0.060
720–800	0.386	0.059
800–900	0.0715	0.067
CO_2 15-μm band		
585–752	718.7	0.448
O_3 9.6-μm band		
1000.0–1006.5	$6.99\cdot10^2$	5.0
1006.5–1013.0	$1.40\cdot10^2$	5.0
1013.0–1019.5	$2.79\cdot10^2$	5.0
1019.5–1026.0	$4.66\cdot10^2$	5.5
1026.0–1032.5	$5.11\cdot10^2$	5.8
1032.5–1039.0	$3.72\cdot10^2$	8.0
1039.0–1045.5	$2.57\cdot10^2$	6.1
1045.5–1052.0	$6.05\cdot10^2$	8.4
1052.0–1058.5	$7.69\cdot10^2$	8.3
1058.5–1065.0	$2.79\cdot10^2$	6.7
H_2O 6.3-μm band		
1,200–1,350	25.65	0.089
1,350–1,450	134.4	0.230
1,450–1,550	632.9	0.320
1,550–1,650	331.2	0.296
1,650–1,750	434.1	0.452
1,750–1,850	136.0	0.359
1,850–1,950	35.65	0.165
1,950–2,050	9.015	0.104
2,050–2,200	1.529	0.116

4.4 Practice 3

4.4.1 Objectives

1. To study the methodology of calculation transmission functions basing on using different models of absorption bands.
2. To calculate with the computer and to study the spectral dependence of the transmission function in IR spectral interval.

4.4.2 Software and Set of Input Parameters

1. Tables containing parameters of the statistical model.
2. Data of gaseous absorbing components content in the atmosphere.

4.4.3 Test Questions

1. What are the assumptions concerning location and intensity of spectral absorption lines used in the considered models of absorption bands?
2. What is the square root law?

4.4.4 Sequential Steps of the Exercise Implementation

1. To study theoretical bases of gaseous absorption lines forming in the atmosphere.
2. To specify initial data for calculating transmission functions in IR interval.
3. To compile the program for PC that implements the calculation of transmission functions in the IR wavelength range on the basis of the teaching model of the absorption band.
4. Build with the help of packages EXCEL, SURFER or TABLECURVE, needed graphs to the report showing the dependence of the transmission as a function of the parameters of the model, and integral content of the absorbing gas.

4.4.5 Requirements to the Report

Compile a concise report reflecting the principal stages, the obtained results in the form of tables and graphs, as well as the basic conclusions. Present the text drawn up by the program to calculate the transmission function in the IR wavelength range based on the absorption band model given by the teacher.

Chapter 5
Calculation of the Intensity of Self Heat Radiation of the System "Surface-Atmosphere"

Abstract The radiative transfer equation for direct heat radiation is solved, that is valid for clear atmosphere in IR spectral region. The cloud cover is taking into account phenomenologically in calculation self heat radiation of the surface.

5.1 Concise Theory

The solution of the radiation transfer Eq. 1.17 by ***ignoring the scattering process*** (that is valid in clear atmosphere for heat radiation in the long-wave spectral ranges ($\lambda > 3$ μm)) and taking into account the definition of *the transmission function* $P(v,x,y)$ according to the Eq. 1.22 that characterized the direct radiation looks as follows:

$$J(v,p_t) = \varepsilon(v)B[v,T_s]P(v,p_S,p_t) + \int\limits_{p_s}^{p_t} B[v,T(x)]\frac{\partial P(v,x,p_S)}{\partial x}dx$$
$$+ [1-\varepsilon(v)]P(v,p_S,p_t)\int\limits_{p_t}^{p_s} B[v,T(x)]\frac{\partial P(v,x,p_S)}{\partial x}dx, \tag{5.1}$$

Here $J(v,p_t)$ is the intensity of the outgoing radiation at the wave number v, at the level with the atmospheric pressure p_t for nadir scheme of observation (viewing angle $\theta = 0$); $\varepsilon(v)$ is the emissivity of the surface; $B[v,T]$ is the Planck's function for the blackbody radiation at the wave number v and temperature T; $P(v,x,y)$ is the transmission function of the monochromatic radiation from the level with pressure y till the level with pressure x, which is the integration variable. It is fixed for outgoing radiation $y = p_t$ at the level of the atmosphere top with satellite scheme of observations, for downward radiation $y = p_S$, where p_S, is the atmospheric pressure at the surface level. T_S is the temperature of the surface. It is to point out that the

I. Melnikova et al., *Remote Sensing of the Environment and Radiation Transfer*,
DOI 10.1007/978-3-642-14899-6_5, © Springer-Verlag Berlin Heidelberg 2012

surface emissivity is below one, and there is a leap of the temperature at the surface: $T_S \neq T(p_t)$. As it follows from the analysis of the Eq. 5.1, the intensity of the outgoing radiation is defined by three items:

- the radiation formed by the surface;
- the radiation formed by the atmosphere;
- the downward radiation formed by the atmosphere and reflected by the surface at the level with the pressure p_t .

We will assume that the surface albedo equals to zero then the 3rd item in the Eq. 5.1 is zero also. Hence, the Eq. 5.1 looks as below:

$$J(v,p_t) = \varepsilon(v)B[v,T_S]P(v,p_S,p_t) - \int_{p_t}^{p_s} B[v,T(x)]\frac{\partial P(v,x,p_t)}{\partial x}dx. \quad (5.2)$$

5.2 Transmission Function

The monochromatic transmission function might be written with the assumption that the radiation absorption at the wave number v is attributable to only one gaseous component of the atmosphere:

$$P(v,p_1,p_2) = \exp\left[-\frac{1}{g}\int_{p_1}^{p_2} k(v,x)q(x)dx\right] p_1 < p_2 \quad (5.3)$$

$$P(v,p_2,p_1) = \exp\left[-\frac{1}{g}\int_{p_2}^{p_1} k(v,x)q(x)dx\right] p_2 < p_1$$

where: g is the acceleration due to gravity at the surface of the earth; $k(v, x)$ is the absorption coefficient at the level with pressure x; $q(x)$ is the specific content of absorbing gaseous component at the level with pressure x.

Values of the emissivity $\varepsilon(v)$ for certain surfaces are in the Table 5.1.

The intensity of outgoing in space radiation in cloudy atmosphere $J_c(v,p_t)$ is characterized by Eq. 5.2, with following changes:

- the atmospheric pressure at the surface level p_S, is replaced by pressure at the cloud top p_c;
- the surface emissivity $\varepsilon(v)$ is replaced by the emissivity at the cloud top $\varepsilon_c(v)$;
- the surface temperature T_S is replaced by the cloud top temperature T_c .

Table 5.1 The emissivity of the surface different types within IR ranges (8–12 μm)

Type of the surface	The emissivity ε
Dry sand with small grains	0.949
Wet sand with small grains	0.962
Dry sand with large grains	0.914
Wet sand with large grains	0.936
Ice	0.980
Clean water	0.993
Quartz	0.712
Granite	0.815
Basalt	0.904
Dry peat	0.970
Wet peat	0.983
Dry sandy loam soil	0.954
Wet sandy loam soil	0.968
Conifer needles	0.971
Dry snow (t =−2.5°C)	0.996
Wet snow	0.997
Dirty snow	0.969
Fresh snow	0.986

In case of a partial cloud cover of the instrument's field of view (cloud part is N_c) the outgoing radiation intensity $J_a(\nu,p_t)$ at the wave number ν is defined by the following:

$$J_a(\nu,p_t) = N_c \cdot J_c(\nu,p_c) + (1 - N_c) \cdot J(\nu,p_s) \tag{5.4}$$

The calculation of the integrals in expressions (5.1) and (5.2) uses numerical approaches:

- the method spline-interpolation of the temperature profile at the same levels, where the transmission function is fixed;
- the method of rectangles for the calculation of the definite integrals:

$$\int_{\varphi(a)}^{\varphi(b)} f(x)d\varphi(x) = \sum_{i=2}^{n} f(x_{i+1/2})\Delta\varphi_i, \tag{5.5}$$

where

$$d\varphi(x) = \frac{\partial\varphi(x)}{\partial x}dx; \quad h = \frac{b-a}{n}, \quad x_i = ih, \quad 1 \le i \le n;$$

$$x_{i-\frac{1}{2}} = \frac{x_{i-1} + x_i}{2}, \quad 2 \le i \le n; \quad \Delta\varphi_i = \left[\varphi\left(x_{i-\frac{1}{2}+1}\right) - \varphi\left(x_{i-\frac{1}{2}}\right)\right], \quad 2 \le i \le n,$$

5.3 Practice 4

5.3.1 Objectives

To study

- specific features of the formation of the heat radiation in cases of clear and cloudy atmosphere and with partly cloud cover of the instrument viewing field;
- the redistribution of the surface yield forming the outgoing radiation within spectral channels with different values of radiation absorption;
- the outgoing radiation dependence on the cloud and surface emissivity, cloud top altitude, degrees of the surface radiative heating (cooling) within different spectral channels.

5.3.2 Software and Set of Input Parameters

1. Computer programs "V_AERDAT.exe", "V_TRANSM.exe" and "RAD_IR.exe" (or "RADNIMB.exe").
2. Tables of results of the complex temperature-wind aerological sounding whose data are used with the program "V_AERDAT.exe" at the first step for creating the input file for further calculation. It is possible to use the ready input files at the address *"\dz-2006\student\"*. The ready file containing values of the transmission function for the radiometer at the platform « *Nimbus* » might be used also with the computer program "RADNIMB.exe" and file *"aer.dat"*.

5.3.3 Test Questions

1. What factors influence the difference in the outgoing intensity values at different wave numbers ν?
2. What does the extent to which the outgoing intensity is affected by the surface emissivity depends on?
5. At which wave number ν used in the calculations is the cloud impact on the outgoing intensity maximum?
4. At which wave numbers ν used in the calculations is the cloud impact on the outgoing intensity maximum?
5. Why is the brightness temperature of the outgoing radiation formed by the system atmosphere-surface not equal to the sum of surface and atmosphere brightness temperatures?
6. In a case, where the isothermal atmosphere temperature is equal to the surface temperature $T(x) = const = T(p_S)$ and the surface emissivity $\varepsilon(\nu)$ is equal to a unit. What is the brightness temperature of the outgoing radiation equal to?

5.3.4 Sequential Steps of the Exercise Implementation

1. To study the theory using additional books, pointed in references list.
2. To use the prepared file "*aer.dat*" containing temperature and pressure profile as initial It is possible to create files containing data of aerological sounding with the program "V_AERDAT.exe". The file is automatically kept in the directory « *\dz-2006\student* » with name "*aer.dat*".
5. To create the file containing values of the transmission function with computer program "V_TRANSM.exe" with name "*Vtransm.dat*" in the same directory as the computer program is. The transmission functions in five spectral channels at 100 levels in the atmosphere ($z = 0 \div 50$ km) are simulated basing on the following equation (an analogue of formulas (5.3)):

$$P\left(v_i,\, z = 50\,\text{km},\; z_j\right) = \exp\left[-k(v_i)\; u(z_j)\right], \tag{5.6}$$

where $z_j = (n - j)/2,\; j = 1,\ldots,101,\; n = 101;\; k(v_i) = 0.005 + (i + 0.05)^{0.8}$, $i = 1, \ldots, 5$;

$u(z_j) = (j/2-1)^{(0.8-0.001\,*\,n)}$.

4. To try calculations with the computer program "RAD_IR" for visualizing all variants for analyzing special features of forming self heat radiation of the system surface-atmosphere in clear and cloudy cases, and different altitudes of cloud top.
5. To prepare the plan of accomplishing numerical experiments with program "RAD_IR.exe".
6. To fulfill whole set of calculations with the program "RAD_IR.exe".
7. To analyze basing on data obtained:

   – the redistribution of the surface and atmosphere contribution at different spectral channels with different radiation absorption in case of clear atmosphere;
   – the dependence of outgoing radiation at different spectral channels with different radiation absorption on the surface emissivity and surface radiation heating (cooling);
   – the cloud impact (altitude and emissivity of cloud top, part of cloud cover of the instrument viewing field) on the outgoing radiation within spectral channels with different radiation absorption.

5.3.5 Requirements to the Report

Prepare the final brief report with elements of theory, resulting pictures, and conclusions that reflect main stages of the work.

Chapter 6
Construction and Operation of the Automated One-Channel IR-Radiometer

Abstract Main features of IR-radiometers, testing and calibration of instrument are considered. Detailed description of all procedures for employment of the IR-radiometer is presented.

6.1 Concise Theory

Mastering space that began in 1957 opens a new era for many research fields. Firstly, the view to the Earth from outside has been possible. A lot of objects were of interest for researches: oceans, continents, glaciers, rivers, and cloudiness. The development of the space technique is aimed to obtaining the maximum of information about geophysical covers and phenomena.

Most of satellites at the orbit are destined for researches (some have several purposes), accomplish investigation of Earth's spheres (hydrosphere, atmosphere, lithosphere, biosphere), and their interaction in global scales. The information source for instruments at satellite platforms is the radiation reflected (shortwave solar) or emitted (heat self radiation) by these spheres. The radiation transformation provoked by different objects at the Earth surface or atmosphere might be treated for these objects properties retrieval.

At first stage observations were similar to photographing and comparing view of objects in different time moments. Natural resources, the dynamics of their variations, anthropogenic impacts and so are were of scientific interest. However, the necessity of obtaining the information concerning physical characteristics arises. These characteristics are as follows:

- the temperature and humidity of the atmospheric air;
- the temperature of the ground and ocean surface;
- determination specific chemical elements in the Earth core;
- the salinity and waving of water surface;
- the altitude of cloud top, ice and water amount of cloud, direction and velocity of cloud shift;

I. Melnikova et al., *Remote Sensing of the Environment and Radiation Transfer*,
DOI 10.1007/978-3-642-14899-6_6, © Springer-Verlag Berlin Heidelberg 2012

– the origin of dangerous atmospheric phenomena (typhoons, squall, tornados, heavy rains, hail and thunderstorm clouds).

Contemporary instruments for remote sensing of the atmosphere and surface are very rich (multi channel radiometers and spectrometers) and registry of the electro-magnetic radiation in different spectral intervals. Classification of instruments can be done on the basis of:

– measurements type (instruments for active and passive remote sensing);
– geometry of observations (nadir and limb);
– spectral intervals where electro-magnetic radiation registered (UV, visual, IR, microwave);
– specific features of observations (with space scanning or not, multi angle, etc.);

In ranges of the above classification of the various set of radiometers is elaborated and successfully used at different platforms (aircraft, balloon, rocket, satellite) for registration of outgoing radiation. IR-radiometers, in particular, serve for remote obtaining of the temperature of the surface and cloud top from boards of aircraft, helicopter and meteorological satellites. Most of the satellite radiometers are scanner type instruments because this type provides the main advances of satellite observation: operative looking through wide territories and mapping with high space resolution.

The first IR-scanner has been launched as a part of satellite complex "Nimbus-2" in 1964. The accuracy was not high – errors of the surface temperature retrieval were about several degrees. But the global scales of obtaining the information demonstrated rich potential possibilities of such instruments and gave the important and practical information.

After testing and tuning instruments at "Nimbus" satellites the improved IR-radiometer has been set operational on meteorological satellites (MS) series ITOC/NOAA at the polar orbit. Efficiency of MS and providing regular and simultaneous IR-radiometer data directly to users (data are transmitted in continues regime) making for wide satellite data assimilation in geophysics.

Launching the geostationary satellite "Meteosat-1" was the next step in development of satellite IR-radiometry. Nowadays several geostationary MS with IR-radiometers at the board simultaneously function at the orbit. Geostationary MS provide the improved data recurrence comparing with MS at polar orbits and characterized with the longer life expectancy. But the space resolution and accuracy of registration IR radiation are lower than for polar orbit satellites.

The MS of the fourth generation "TYROS-N" began to operate in 1978. Its improved version was the basic element of USA space system till the end of 80th. The radiometer AVHRR has been put at this MS and the same instrument was used at the board of NOAA NROSS satellite for oceanic researches.

With creating improved receivers of the IR-radiation the IR-radiometry IR-spectroscopy techniques are brought closer, that leads elaborating new improved approaches of satellite data processing. Thermocouples and bolometers (which are of high sensitivity and offer freedom of cooling) were used in first generation radiometers. However, they possess relatively high (for the satellite observations specificity) time constant of order tens and hundreds of milliseconds that prevents high space resolution. It is mentioned for comparison that the registration time of one resolution element by the radiometer AVHRR is 0.25 ms. Contemporary IR-radiometry derives success in improvements from creating photoelectric receivers with high spectral sensitivity (in spectral ranges 8–14 μm) Photoresistors and photodiodes on the base of *HgColTe* are in considerably use. The maximum of their spectral sensitivity is at 10 μm and the time constant ranges within 50–500 ns with receiver cooling till 77 K. The significant advance of photoelectric receivers is the possibility organizing photocells to lines and matrixes that allows avoiding complicated optics-mechanics scanning systems in radiometers. New receivers call putting new technologic problems for creators of satellite instruments: special microcryogenic systems for receivers cooling with small mass and low energy consumption; the necessity of fabricating receivers similar over their parameters (that is important for multi-element receivers).

Improving the methodology of multichannel observations (elucidating the optimal spectral intervals and their number) increase the remote observations as well. Nowadays there are not principal difficulties in creating multichannel radiometers. A different situation arises with perspective constructions (heterodyne radiometers, IR-spectrometers) that can provide the qualitative leap in accuracy improvement of radiometers. The stable and highly coherent radiation sources are needed for creating and using heterodyne radiometers that call cryogen system for satellite IR-spectrometers.

Graduation of satellite radiometers is of most important stage in the instrument preparation because it defines the results accuracy of meteorological parameters measurements. It is clear that graduation of satellite instruments is to be accomplished in similar to operational conditions at satellite board.

However, the extended and careful graduating preliminary program of the instrument is not enough for necessary accuracy of radiometric observation. In particular 1–2 months are needed after instrument launching at the MS board for stable running radiometer in operating regime ratings. For example, the radiometer *HCMR* regime ratings characterized by the receiver temperature 150 K, but first 70 days after launching the temperature exceeds 200 K and decreases to 0.14 K per day during 30–50 days.

The orbital calibration is accomplished during all period of instrument's operation that regulates by following factors:

- "ageing" electronics that leads variations of the instrument sensitivity;
- changing conditions of solar heating the satellite and instrument;
- Possibility of ice deposition at cooled instrument elements because of degassing and condensing processes.

Thus satellite radiometers demand continuous orbital calibrating during operational cycle at MS board for expeditious correction of graduating characteristics. From the experience the orbital calibration is accomplished with both the reference radiation of the space and boarding imitating of the blackbody. The absolute calibration only with the blackbody is not sufficient.

6.2 The Function and Observing Conditions of the Automated One-Channel IR-Radiometer

This practice concerns studying, graduating and calibrating the IR-radiometer. Certain details of construction are special cases. However, considered principles of operation and construction are inherent features in general for IR-radiometers. Thus let's consider the interaction between separate units of the instrument without concrete specification of elements.

The automated one-channel IR-radiometer used in this teaching practice is fabricated for remote measurement of the temperature of water surface (or other surface with known emissivity) from the aircraft or helicopter board. The instrument is aimed to nadir observation in atmospheric window 8–13 μm. Space scanning is done by observational platform moving.

The instrument is compiled from two parts: the optical unit and the unit for processing signal. The optical unit can be put in both hermetic sealed and non-hermetic compartment of aircraft or helicopter. The optical axis is vertical for nadir observation.

Operating the instrument allows following variants:

(a) registering observational results in digital form with indicators consisting from light-emitting diodes;
(b) outputting the signal in digital form to computer (this option is used in laboratory module);
(c) outputting the signal in analog form (1, . . ., 5 V) for registering with the analog recorder.

Normal conditions for the processing unit are follows:

(a) the air temperature is in ranges	288–298 K (15–25°C);
(b) the air relative humidity is in ranges	15–65% for pointed temperatures;
(c) the atmospheric pressure is in ranges	96–104 kPa (720–780 mmHg);
(d) the supply voltage is	215.6–224.4 V and the frequency is 50 Hz;

Working conditions for the processing unit are follows:

(a) the air temperature is in ranges	278–313 K (5–40°C);
(b) the atmospheric pressure is in ranges	96–104 kPa (720–780 mmHg);

(c) the supply is possible from both the alternating current with the frequency 49.5–50.5 Hz and voltage in ranges 198–244 V or the direct current with the voltage in ranges 24–30 V.

Working conditions for optical unit (with turned on thermostats) are follows:

(a) the air temperature is in ranges	253–303 K (−20 to +30)°C;
(b) the air relative humidity is in ranges	98% for temperature 303 K (30°C);
(c) the atmospheric pressure is in ranges	33–104 kPa (250–780 mmHg).

Limiting conditions for transportation the instrument are:

(a) the air temperature is in ranges	253–333 K (−20 to +60)°C;
(b) the air relative humidity is in ranges	99% for temperature 303 K (30°C);

6.3 Technical Parameters of the Automated One-Channel IR-Radiometer

The range of measured temperatures of the water surface:	0–40°C
The instrument error is not more than:	0.3 K
The resolution at the digital output	0.1 K
The viewing indicator is not more than	1:20
The temperature of thermostats	35°C
The time constant of the integrator	2 s
The spectral interval,	8–13 μm
The signal output is analog	1–5 V
The instrument size are, mm:	
optical unit not more than	80 × 180 × 260 mm
processing unit not more than	260 × 150 × 260
The instrument mass is not more than	8 kg
The length of cable between optical and processing units is not more than	20 m

6.4 General Instructions on Exploring the IR-Radiometer

Firstly, it should carefully familiarize with the technical description and instruction on exploring of the instrument before the operation begins.

Governing elements of the IR-radiometer are at the front signal processing panel.

The first assembly connector (farthest to the left) is destined for connecting the cable from the optical unit (marked "optical unit"). The second assembly connector (without marks) is for output analog and digit signals. The third connector is for supply cable and marked "power".

Caution should be exercised in handling the instrument. One should be careful while operating, calibrating or repairing the instrument and keep off from touching energized elements because the alternate voltage 220 V is in the processing unit.

6.5 Preparing the IR-Radiometer to Operation

1. Study the function of every governing element.
2. Connect with the cable the optical and processing units and switch the power cable.
3. Switch the power cable to the power source.

4. Switch the toggle switch "power" to the lower position.
5. Open the protective cover from the front panel of optical unit.

6.5.1 The Order of Measurement

After switching the toggle switch “power” three digits and two commas are lightning The instrument is to heated during 5–15 m for stabilizing regime of thermostat (disappearing of the second comma at lightning digit panel is the signal of the operative regime).

The optical unit is to be positioned at operational place (move the platform with the optical unit to the place, which is above the surface point with the desired temperature). After 2 s the light indicators demonstrate the water surface temperature (°C).

6.5.2 Testing the IR-Radiometer

During testing the laboratory glass mercury thermometer with the least graduation 0.1°C and scale limits 0–50°C is to be used as a reference thermometer.

Environmental conditions during testing:

- the air temperature	10–25°C
- the air relative humidity	50–70%
- the atmospheric pressure	96–104 kPa
- the power voltage	200–240 V
- the power frequency	50 Hz
or direct voltage	24–36 V

It is necessary heating the radiometer during 15 m after switching. The reference surface is the special dish with fresh water. It is possible to use the outer blackbody radiator for testing the signal dispersion.

6.5.3 The Calibration Procedure

1. To measure and memorize the temperature of the upper 2 cm layer of water with continuous stirring in the dish. Take in mind that the time constant of the laboratory thermometer is about 3 m.
2. To direct the optical unit to the interested surface and to continue stirring water.
3. To record the radiometer result.

4. To repeat the procedure above the desired point of the water surface ten times with recording results of radiometer and reference thermometer.
5. To repeat the procedure at other 10–15 points of the surface for other surface temperatures in ranges 0–30°C.
6. To correct the tuning IR-radiometer if temperature recording by radiometer and reference thermometer distinguish more than 0.3 K.

Remark. Recording by radiometer might exceed thermometer recording by 0.5–0.8°C in the temperature range 0–9°C, because of the warm skin film on the surface that is in contact with the warm laboratory air. By contrast in the temperature range 16–30°C radiometer recording might be lower by 0.5–1 K than thermometer recording owing to surface skin film heat losses through the evaporation.

The Fig. 6.1 demonstrates the plot of testing results. On the abscissa recordings by IR-radiometer (*Tr*) are plotted and on the ordinate the difference (*dT*) between remote by radiometer (*Tr*) and contact by thermometer (*Ts*) recordings are plotted. The Table 6.1. presents the similar data in tabulated view. The special file "*kalib1.dat*" with calibrating values can be used for quantitative analysis during observation.

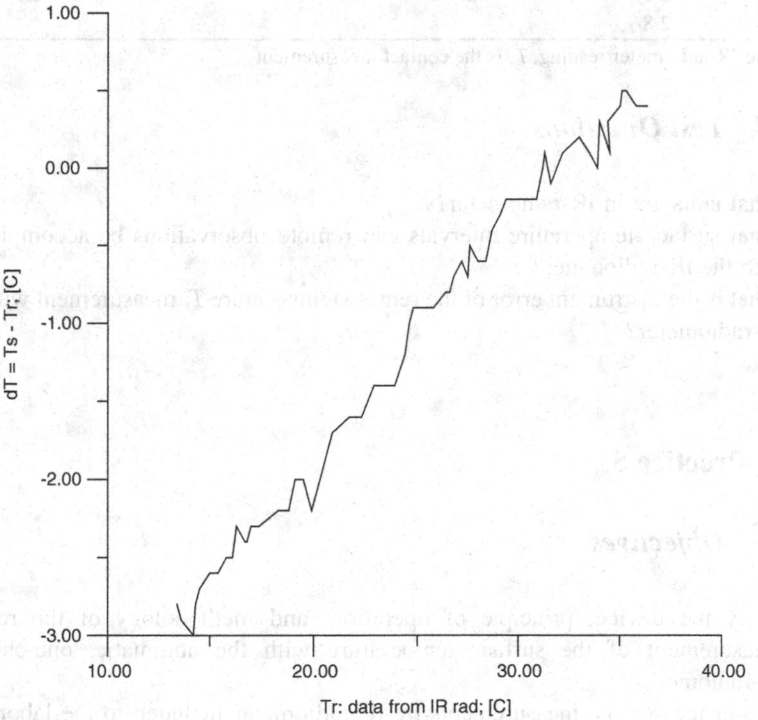

Fig. 6.1 Example of the IR-radiometer calibration

Table 6.1 Results of IR-radiometer calibration

T_r [°C]	$T_s - T_r$ [°C]	T_r [°C]	$T_s - T_r$ [°C]	T_s [°C]	$T_s - T_r$ [°C]
36.3	0.4	35.8	0.4	35.3	0.5
35.1	0.5	35.0	0.4	34.4	0.3
34.5	0.1	34.0	0.3	33.9	0.0
33.5	0.1	33.0	0.2	32.2	0.1
31.6	−0.1	31.3	0.1	31.2	0.0
30.9	−0.2	30.6	−0.2	30.3	−0.2
29.9	−0.2	29.6	−0.2	29.4	−0.2
29.1	−0.3	28.7	−0.4	28.4	−0.6
28.0	−0.6	27.6	−0.5	27.5	−0.7
27.2	−0.6	26.8	−0.7	26.6	−0.8
26.4	−0.8	25.8	−0.9	25.6	−0.9
24.9	−0.9	24.8	−0.9	24.6	−1.0
24.4	−1.2	23.9	−1.4	22.9	−1.4
22.3	−1.6	21.7	−1.6	20.9	−1.7
19.9	−2.2	19.5	−2.0	19.1	−2.0
18.8	−2.2	18.3	−2.2	17.4	−2.3
17.0	−2.3	16.8	−2.4	16.7	−2.4
16.3	−2.3	16.1	−2.5	15.8	−2.5
15.4	−2.6	15.0	−2.6	14.5	−2.7
14.3	−2.8	14.2	−3.0	13.6	−2.9
13.4	−2.8				

T_r is the IR-radiometer reading, T_s is the contact measurement

6.5.4 Test Questions

1. What units are in IR-radiometer?
2. What surface temperature intervals can remote observations be accomplished with the IR-radiometer?
3. What is the instrument error of the remote temperature T_s measurement with the IR-radiometer?

6.6 Practice 5

6.6.1 Objectives

1. Study the device, principle of operation, and methodology of the remote measurement of the surface temperature with the automatic, one-channel IR-radiometer.
2. Master the remote measurements by IR-radiometer included to the laboratory complex "IR-radiometer–Computer".
3. Carry out testing the IR-radiometer.

6.6.2 *Software and Set of Input Parameters*

1. The automated one-channel IR-radiometer, mated with the computer.
2. Thermometers for contact temperature measurements.
3. Computer programs for the surface temperature processing. Files names and paths of needed programs are pointed (given) by the professor.

6.6.3 *Sequential Steps of the Exercise Implementation*

1. Read attentively the section devoted to the physical/theoretical background of this exercise. If necessary, consult the referenced literature.
2. Explain the optical scheme of the operation IR-radiometer using the Fig. 6.1.
3. Master computer programs for resulting data processing with computer.
4. Accomplish the IR-radiometer testing and using dishes with water at different temperatures. Estimate average and maximal observational errors within given temperature ranges.
5. Prepare the report containing concise description of the radiometer and principles of operation and results of testing radiometer with estimating observational errors.

Chapter 7
Remote Measurement of the Surface Temperature Field with the Automated One-Channel IR-Radiometer

Abstract Basic formulas are considered for the surface emissivity and temperature retrieval from the radiation observation. The sequential steps for the Practice implementation are presented.

7.1 Concise Theory

The intensity of self heat radiation of the system "surface-atmosphere" is described by the Eq. 8.6. Let's repeat it here:

$$J_\nu^{\uparrow} = \varepsilon_\nu B[\nu, T_s] P_\nu^{(s)} + B\left[\nu, T_\nu^a\right]\left\{1 - P_\nu^{(s)}\right\}, \tag{7.1}$$

where

$J_\nu^{\uparrow}$ is the intensity of self heat radiation at the wave number ν;
ε_ν is the surface emissivity at the wave number ν;
$B[\nu, T_s]$ is Planck's function at the wave number ν and temperature T_s;
$P_\nu^{(s)}$ is the transmission function of whole atmosphere at the wave number ν;
T_ν^a is the "effective temperature" of the atmosphere at the wave number ν.

It is possible to assume $P_\nu^{(s)} \approx 1$ considering that observations are in atmosphere window and the distance between radiometer and surface is about 1 m. Then

$$J_\nu^{\uparrow} \approx \varepsilon_\nu B[\nu, T_s]. \tag{7.2}$$

The relation (7.2) might be used for the surface temperature T_s retrieval after substituting the expression of Planck's function (2.1) if the radiometer records the intensity $J_\nu^{\uparrow}$ as such. Let's remember the expression for the Planck's function:

I. Melnikova et al., *Remote Sensing of the Environment and Radiation Transfer*, DOI 10.1007/978-3-642-14899-6_7,

$$B\left[\nu, T_s\right] = \frac{a\nu^3}{\exp\left(\frac{b\nu}{T_s}\right) - 1}, \tag{7.3}$$

where a = 1.1909 · 10^{-5}, b = 1.438786, dimensions of the wave number and temperature are: [ν] = см^{-1}, and [*T*] = ^{0}K.

Actually, by taking into account for the expression

$$J_\nu^\uparrow = \varepsilon_\nu \frac{a\nu^3}{\exp\left(\frac{b\nu}{T_s}\right) - 1}, \tag{7.4}$$

The expression for obtaining surface temperature is obtained after elementary transformations

$$T_s = \frac{b\nu}{\ln\left[\varepsilon_\nu \frac{a\nu^3}{J_\nu^\uparrow} + 1\right]}. \tag{7.5}$$

The Eq. 7.4 together with direct measurements of the surface temperature T_s might be evidently used for retrieving the surface emissivity ε_ν:

$$\varepsilon_\nu = J_\nu^\uparrow \frac{\exp\left(\frac{b\nu}{T_s}\right) - 1}{a\nu^3}. \tag{7.6}$$

However, the above relations do not take into account for special features of the radiometer calibration, they can't be directly used. The instrument calibration has been fulfilled for the water surface (with the emissivity $\varepsilon_\nu = 0.993$) only, the calibration curve memorized by computer is not suitable for another surface, and the temperature recorded by radiometer does not correspond to real one. Thus, radiometer's reading is necessary to correct, if the other surfaces are under consideration for temperature and emissivity retrieval.

Let the methodology for recounting the radiometer reading consider when accomplishing observation.

7.2 Determination of the Surface Emissivity

Input data for determining the emissivity ε_ν are:

- The surface temperature T_s measured with contact method;
- Radiometer recordings T_r.

IR-radiometer records the self heat radiation intensity above water surface

$$J_\nu^\uparrow = \varepsilon_\nu^{(w)} B[\nu, T_r], \tag{7.7}$$

and the following value lights at the radiometer panel:

$$T_r = \frac{bv}{\ln\left[\varepsilon_v^{(w)} \frac{av^3}{J_v^{\uparrow}} + 1\right]}, \tag{7.8}$$

which is not equal to the real temperature value because of inequality emissivity of water surface to emissivity of other studied surface $\varepsilon_v{}^{(w)} \neq \varepsilon_v$.

Actually we have

$$J_v^{\uparrow} = \varepsilon_v B[v, T_s], \tag{7.9}$$

where ε_v and T_s are the emissivity and temperature of studied surface. It is evident that:

$$\varepsilon_v^{(w)} B[v, T_r] = \varepsilon_v B[v, T_s], \tag{7.10}$$

where T_r is the radiometer reading and T_s is the surface temperature. Then the following relation is obtained with assuming known values $\varepsilon_v^{(w)}$, T_r and T_s:

$$E_v = \varepsilon_v^{(w)} \frac{B[v, T_r]}{B[v, T_s]} \tag{7.11}$$

With substituting Planck's function expression to the Eq. 7.10 the formula for calculating the real emissivity value is derived:

$$\varepsilon_v = \varepsilon_v^{(w)} \frac{\exp\left(\frac{bv}{T_s}\right) + 1}{\exp\left(\frac{bv}{T_r}\right) + 1}, \tag{7.12}$$

where v =1,000 cm^{-1} (the centre of radiometer filter transmission); $\varepsilon_v^{(w)}$=0.993; b = 1.438786 K/cm^{-1}. The Eq. 7.12 is used in the computer program « RAD.exe » for the determination of emissivity ε_v.

7.3 Remote Measurement of the Surface Temperature

The direct radiometer reading is used for remote measurement of water surface temperature.

Radiometer reading is not sufficient for observation above other surfaces (sand, soil, et al.) for surface temperature T_s retrieval. Value of emissivity ε_v of corresponding surface (that are measured independently following the previous section) is necessary. The recalculation radiometer reading T_r to real value T_s is done with Eqs. 7.7 and 7.9. After equating right parts of the equations and substituting Planck's formula the following is obtained:

$$\frac{\varepsilon_\nu a\nu^3}{\exp\left(\frac{b\nu}{T_s}\right)+1}=\frac{\varepsilon_\nu^{(w)}a\nu^3}{\exp\left(\frac{b\nu}{T_r}\right)+1}. \tag{7.13}$$

Solution of the Eq. 7.13 relatively T_s yields:

$$T_s=\frac{b\nu}{\ln\left\{\frac{\varepsilon_\nu}{\varepsilon_\nu^{(w)}}\left[\exp\left(\frac{b\nu}{T_r}\right)+1\right]-1\right\}}. \tag{7.14}$$

The Eq. 7.14 is used for obtaining value T_s with recalculating radiometer reading T_r and values ν =1,000 cm^{-1}, b = 1.438786°K/cm^{-1}, $\varepsilon_\nu^{(w)}=0.993$ in the computer program « RAD.exe ».

7.4 Polynomial Approximation of the Temperature Field Measured with One-Channel Automated IR-Radiometer

The typical situation, when solving meteorological problems, is the obtaining meteorological parameter at nodes of regular network from observational value of the meteorological parameter at arbitrarily posed sites at horizontal plane. The bi-dimension polynomial approximation is one of approaches to the problem solution. In particular the first attempts of meteorological fields analysis has been based on the polynomial approximation. Experience showed that the approach provides acceptable exactness of the analysis with observational sites thickly strewn. However the approach might provoke a significant error for vast space with rare observational sites. Nowadays the spline approximation inspires the interest to the polynomial approximation.

Here the polynomial approximation is used for interpretation of remote data of temperature measurement. The algorithm for calculating coefficients of approximating polynomial is considered.

Let observational results be known for a certain meteorological value H at N points of horizontal plane with coordinates (x_i, y_i), where i = 1, 2, … , N. These values are noted H_i. The totality of values x_i, y_i, and H_i are called table data.

The bi-dimension field of the meteorological value H is defined by the following expression with using polynomial approximation:

$$H(x,y)=\sum_{j=1}^{m}f_jF_j(x,y), \tag{7.15}$$

where $F_j(x,y)$ are known power monomial in coordinates x and y, and f_j are unknown desired coefficients (found on the base of all observational data totality)

Depending on the power of used polynomial *sp* power monomials F_j and parameter m in the Eq. 7.15 are as follows:

- for linear approximation ($sp = 1$) $m = 3$; $F_1 = 1$; $F_2 = x$; $F_3 = y$;
- for square approximation ($sp = 2$) $m = 6$; $F_4 = x\,y$; $F_5 = x^2$; $F_6 = y^2$ (power monomials F_0, F_1 and F_2 are identical to case of the linear approximation);
- for cubic approximation ($sp = 3$) $m = 10$; $F_7 = x^2\,y$; $F_8 = x\,y^2$; $F_9 = x^3$; $F_{10} = y^3$ (power monomials F_0–F_9 are identical to case of the square approximation).

Number of coefficients of approximation polynomial m is linked with its power *sp* with the relation:

$$m = \sum_{i=0}^{sp} (i+1). \tag{7.16}$$

Different methods are possible for determination of coefficients f_j in the Eq. 7.15, which differ by mathematical approaches and totality of used data both real (measured) and a priori (known before an experiment). Here the calculation of coefficients f_j is done in the range of linear theory of the less-square technique for independent and equally accurate observational data. Coefficients f_j are calculated from the demand of minimum by the following value:

$$E(f_1,\ \ldots.\ ,f_m) = \sum_{i=1}^{N} \left[H(x_i, y_i) - H_i\right]^2 , \tag{7.17}$$

where $H(x_i,y_i)$ are values of approximating polynomial defined with the Eq. 7.15 at points with coordinates (x_i,y_i), H_i are measured values of the meteorological parameter at the same points. The value E is the function of m variables f_1–f_m. Every partial derivative with respect to f_1–f_m is equal to zero at point of the corresponding minimum. Presenting derivatives in an explicit form provides the system of m linear algebraic equations for determination m unknown coefficients f_1–f_m:

$$\begin{cases} \dfrac{\partial E}{\partial f_1} = 0 \\ \dfrac{\partial E}{\partial f_2} = 0 \\ \ldots\ldots\ldots\ldots \\ \dfrac{\partial E}{\partial f_m} = 0 \end{cases}. \tag{7.18}$$

The system (7.18) might be presented in matrix form after substituting the Eq. 7.15 to the formula (7.17), differentiating and elementary algebraic transformations:

$$Df = \mu, \tag{7.19}$$

where the ($m \times 1$)-dimensional vector f contains desired coefficients of the approximating polynomial, the m-by-m matrix D contains only coordinates of observational points, the ($m \times 1$)- dimensional vector μ contains coordinates of observational points with measured data. Then the problem solution might be presented as follows:

$$f = D^{-1}\mu\,, \tag{7.20}$$

where D^{-1} is the matrix inverse of matrix D.

Thus, the calculating coefficients of approximating polynomial is reduced to forming the matrix D and vector μ on the basis of available data, Calculating the inverse matrix D^{-1}, and multiplying to the vector μ.

It is convenient to introduce the auxiliary m-by-N matrix A for forming elements of the matrix D and vector μ of the system (7.20), which for polynomials in first and third powers looks as

- for the first power polynomial:

$$\begin{aligned} A(i,1) &= 1; \\ A(i,2) &= x(i); \\ A(i,3) &= y(i); \end{aligned} \tag{7.21}$$

- for the third power polynomial:

$$\begin{aligned} A(i,1) &= 1; \\ A(i,2) &= x(i); \\ A(i,3) &= y(i); \\ A(i,4) &= x(i)\cdot y(i); \\ A(i,5) &= x(i)\cdot x(i); \\ A(i,6) &= y(i)\cdot y(i); \\ A(i,7) &= x(i)\cdot x(i)\cdot y(i); \\ A(i,9) &= x(i)\cdot x(i)\cdot x(i); \\ A(i,10) &= y(i)\cdot y(i)\cdot y(i). \end{aligned} \tag{7.22}$$

The parameter m for the power sp polynomial is defined by the Eq. 7.17 and N is the number of table value of the field. In the similar manner the expressions might be obtained for other powers polynomials.

In addition the $(N \times 1)$-dimension vector T

$$T(i) = H_i, \tag{7.23}$$

that contains all measured values of the desired field. Then the matrix D and vector μ from the Eq. 7.19 might be obtained with the following relations:

$$\mathrm{D} = \mathrm{A} \times \mathrm{A}^t;\ \mu = \mathrm{A} \times \mathrm{T}, \tag{7.24}$$

where the letter "t" denotes transposed matrix.

It is useful normalizing table data (x, y and H) over the interval [−1, +1] for decreasing uncertainty, when calculating polynomial coefficients, according to the following formula:

$$p_n = 2\frac{p - p_{\min}}{p_{\max} - p_{\min}} - 1, \tag{7.25}$$

where p is the initial value, p_n is the normalized value, p_{min} and p_{max} corresponding minimal and maximal values of the desired parameter over all table data totality.

7.5 Control Questions

1. What surface type does the radiometer panel reading indicate?
2. What values are needed for remote measurements of the surface temperature with IR-radiometer?
3. What values are needed for remote measurements of the surface emissivity with IR-radiometer?
4. Do approximate polynomial values equal to temperature values measured with IR-radiometer at sites of remote observation?

7.6 Practice 6

7.6.1 Objectives

1. Study the methodology of accomplishing remote measurement of the surface temperature with automated IR-radiometer.
2. Master the methodology of using IR-radiometer included in the laboratory complex radiometer-computer for accomplishing remote measurements of the surface temperature and emissivity.
3. Accomplish measurements of the emissivity of proposed type of surface and estimate the result's accuracy.

4. Accomplish the remote measurements of the surface temperature field and create the corresponding data file using obtained surface emissivity.
5. Complete the complex processing data of remote measurement of the surface temperature:
   - calculate coefficients of 2-power approximating polynomial from remote measurement of the surface temperature field;
   - Perform the approximation of observational data to regular network;
   - Present the final result in both digital and pseudo color forms;
   - Plot isometric lines of the surface temperature.
6. Coincide direct and remote measurements of the surface temperature.
7. Analyze the water surface pollution impact on remote measurement of the surface temperature and estimate the possibility of the remote indicating polluted areas.

7.6.2 Software and Set of Input Parameters

The automated one-channel IR-radiometer, mated with the computer.

1. Thermometers for contact temperature measurements.
2. Dish with water or other type of surface (sand, soil).
3. Computer programs for the surface temperature processing ("POLETS.exe" and "TXT2ISO.exe"), as well as graphical packages SURFER or EXCEL editors.

7.6.3 Sequential Steps of the Exercise Implementation

1. The sand emissivity ε_v is defined on the basis of remote and contact measurement of the surface temperature with the computer program "RAD.exe" in option « Definition of sand epsilon ».
   It needs the following operations:
   - To choose the item « Definition of sand epsilon » with buttons "↑" and "↓" after running the program "RAD.exe";
   - To input the surface temperature value (measured with contact thermometer during 3 min);
   - To remember the obtained emissivity value ε_{sand}.
2. To accomplish remote measurement of the surface temperature field $T_s(x, y)$ and to process obtained data complexly:
   - To turn on the electric heater (at one corner of the dish) during 3–5 min for creating the temperature heterogeneity of the surface in dish;

- In running program "RAD.exe" (to press the button *"ENTER"* for continuation the program after the emissivity definition) to open "Menu" by pressing the button *"F6"* and to choose the item "measurement of surface temperature field" with help of buttons "↑" and "↓";
- To put the IR-radiometer above new point of the surface with coordinates X,Y (cm)
- After 30 s to register the radiometer reading to data file by pressing the button "Y". Continuation of registering is done by repeating the "Y" button press;
- To repeat measurement for all points;
- After last measurement (pressing the button "Y") to terminate the process by pressing the button *"N"*;
- Then the program process data. After the program's finish the sand temperature field appears at the computer screen. The map might be moved by buttons "↑", "↓", "←", "→", and changed in sizes by buttons "+", "−";
- To analyze results obtained;
- For terminating the program to press the button *"Esc"*, choose the option *"Quit"*, and then press buttons: *"Esc"*, "TERMINATE", *"Y"*.

3. To accomplish the polynomial approximation of remote measurement data of the surface temperature field. To execute the recalculation of temperature at the regular network basing on obtained values of approximating polynomial coefficients and plotting results:

   - Use the file "*Results.dat*" and remember it in your own directory.
   - To calculate approximating polynomial coefficients with helping programs "POLETS.exe", and "TXT2ISO.exe", and inputting name of the file containing T_s (and "file path"), calculate coefficients of polynomial powers 1–4, that sequentially output at the screen and to the result file "....";
   - To analyze the accuracy of the approximation with comparing approximated and contact measured temperatures T_s at points x and y. The mean square, minimal, and maximal deviations are also output at the screen and to the file "*Results.dat*";
   - To execute recording two result files at the regular network. The notations: "*rrr.txt*" and "*ppp.txt*" are added respectively to the name of resulting file for non-normalized and normalized values T_s in the grid correspondingly, when the file recorded on a network drive;
   - To map "EXCEL" editor and to analyze mapping of surface temperature fields in color gradation;
   - To plot temperature isolines with program "TXT2ISO" and EXCEL editor. Files "*.txt*", created with program "POLETS.exe" are used;
   - To plot results using EXCEL editor and compile a report using the Word text editor.

4. To accomplish remote measurement of the temperature of water surface in two dishes: with pure water and water polluted with oil. To measure the temperature

with contact thermometer in dish #1 (pure water) after stirring water and compare with radiometer reading.

5. To measure the temperature with contact thermometer in dish #2 (water with oil film) and define the temperature difference between pure and polluted water surfaces. Estimate the possibility of indicating the oil pollution of the water surface from temperature difference.
6. To prepare the report in Word text editor.

7.6.4 Requirements to the Report

Report must contain the following:

1. Concise description of the methodology of obtaining sand surface emissivity with IR-radiometer.
2. Results of the complex processing data of the surface temperature remote measurements.
3. Results of possibility of remote indication of oil spots on the water surface.

Chapter 8
Study of Depending the Uncertainty of the Remote Surface Temperature Retrieval on the Initial Parameters Exactness

Abstract The inverse problem of the surface temperature retrieval is formulated. The influence of the initial parameters uncertainties to the resulting temperature is considered.

8.1 Remote Retrieval of the Surface Temperature

Remember the expression (2.9) for the outgoing intensity formed by the surface with ignoring the atmosphere:

$$J_\nu^\uparrow = \varepsilon_\nu B_\nu(T_s), \tag{8.1}$$

where T_S is the surface temperature; B_ν is the Planck's function; ε_ν is the surface emissivity.

Taking into account only the atmospheric absorption and neglecting scattering (that is valid in IR spectral region $\lambda > 3$ μm without cloud) the relation (8.1) transformed to:

$$J_\nu^\uparrow = \varepsilon_\nu B[\nu, T_s] P_\nu(p_S) + \int_{p_s}^{p_t} B[\nu, T(x)] dP_\nu(x). \tag{8.2}$$

Here $P_\nu(x)$ is the transmission function at the wave number ν between levels in the atmosphere with pressure p_t (the atmosphere top) and x (the integrating variable); p_S is the pressure at the surface level.

The following expression is valid for the 2nd item in the Eq. 8.2: according to mean value theorem:

$$\int_{P_s}^{P_t} B[\nu, T(x)] dP_\nu(x) = B\left[\nu, \tilde{T}_\nu\right][1 - P_\nu(p_s)], \tag{8.3}$$

I. Melnikova et al., *Remote Sensing of the Environment and Radiation Transfer*, DOI 10.1007/978-3-642-14899-6_8, © Springer-Verlag Berlin Heidelberg 2012

where $\tilde{T}_\nu$ is certain, generally speaking unknown, temperature value ("mean" temperature of the atmosphere at the wave number ν).

The trivial relation $P_\nu(p_t) = 1$ is used for deriving the Eq. 8.3. The value $\tilde{T}_\nu$ is found from the Eq. 8.2 solution:

$$\tilde{T}_\nu = \frac{b\nu}{\ln\left[1 + \frac{a\nu^2}{D_\nu}\right]}, \tag{8.4}$$

where

$$D_\nu = \frac{1}{1 - P_\nu(p_s)} \int_{p_s}^{p_t} B[\nu, T(x)]dP_\nu(x). \tag{8.5}$$

The Eq. 8.2 might be transformed using the Eq. 8.3 to:

$$J_\nu^\uparrow = \varepsilon_\nu B[\nu, T_s]P_\nu(p_s) + B\left[\nu, \tilde{T}_\nu\right][1 - P_\nu(p_s)]. \tag{8.6}$$

After substituting the Planck's function presentation (Eq. 2.1) to the Eq. 8.6 and solution relative the variable T_S, the formula for the surface temperature retrieval from observed outgoing intensity might be obtained. This formula is valid within the spectral ranges where the absorption is weak (quasi-transparent) that called *window*:

$$T_s = \frac{b\nu}{\ln\left[1 + \varepsilon_\nu P_\nu(p_s) a\frac{\nu^3}{K_\nu}\right]}, \tag{8.7}$$

where

$$K_\nu = J_\nu - B\left[\nu, \tilde{T}_\nu\right](1 - P_\nu(p_s)). \tag{8.8}$$

Thus Eqs. 8.7 and 8.8 provide obtaining the temperature T_s from observational data J_ν if parameters $\varepsilon(\nu)$, $P_\nu(p_s)$ and $\tilde{T}_\nu$ are known.

8.2 Analytical Approaches to the Estimation of Uncertainty of the Surface Temperature T_s Retrieval

The Eq. 8.7 is considered as a function of four variables for estimating the uncertainty of the temperature retrieval:

$$T_s = T_s(x_1, ..., x_4), \tag{8.9}$$

where $x_1 = \varepsilon_v$, $x_2 = J_v$, $x_3 = P_v(p_s)$, $x_4 = \tilde{T}_v$.

Then using the function $T_S(x_1,\ldots,x_4)$ to the Tailor expansion and ignoring items containing dx_1 is larger in power then unit the expression is derived for estimating the absolute error of the temperature T_S:

$$|\Delta T_s| = \sum_{i=1}^{4} \left| \frac{\partial T_s}{\partial x_i} \right|_{x_i=\bar{x}_i, \quad i=1,\ldots,4} \cdot |\Delta x_i|, \tag{8.10}$$

where Δx_i is the uncertainty of the parameter x_i measurements; $\bar{x}_i$ is the exact value of the i-th parameter.

Calculation of corresponded derivatives leads to the relation:

$$\begin{aligned} |\Delta T_s| &= L_v P_v(p_s) T_s |\Delta \varepsilon_v| + \frac{L_v}{K_v} \varepsilon_v P_v(p_s) T_s |\Delta J_v| \\ &+ L_v \varepsilon_v \left[1 - P_v(p_s) \frac{B\left[v, \tilde{T}_v\right]}{K_v} \right] T |\Delta P_v(p_s)| \\ &+ \varepsilon_v P_v(p_s) [1 - P_v(p_s)] \exp\left(\frac{bv}{\tilde{T}_v} \right) B\left[v, \tilde{T}_v\right]^2 \frac{T_s^2 |\Delta \tilde{T}_v|}{D_v K_v^2 \tilde{T}_v^2}. \end{aligned} \tag{8.11}$$

Following notations (symbols) are introduced in the Eq. 8.11:

$$D_v = \left(1 + \varepsilon_v P_v(p_s) \frac{a v^3}{K_v} \right), \tag{8.12}$$

$$L_v = \frac{a v^3}{K_v D_v \ln(D_v)}. \tag{8.13}$$

Different approaches might be used for estimating the value $|\Delta T_s|$.

The first approach. It is necessary to specify for estimating the influence of a certain parameter α on the value $|\Delta T_s|$ the exact value of the parameter $\alpha = \alpha_0$ and the measured value containing an uncertainty (error)

$$\alpha_1 = \alpha_0 - \Delta \alpha_0 \tag{8.14}$$

After calculation $T_s(\alpha_0)$ and $T_s(\alpha_1)$ the desired uncertainty $|\Delta T_s|$ might be found from the relation:

$$|\Delta T| \, |_{\alpha_0, \Delta\alpha} = |T_s(\alpha_1) - T_s(\alpha_0)|. \tag{8.15}$$

Here the value T_s depends on four parameters, thus corresponded estimations might be obtained with varying one parameter and fixing another three or considering variations of any of its combinations.

The second approach. The estimation of $|\Delta T_s|$ is possible to obtain with Eqs. 8.11, 8.12 and 8.13. Calculating every item gives the contribution of every parameter to the uncertainty $|\Delta T_s|$.

The third approach is based on using the Monte-Carlo method. Consider this approach more detailed.

8.3 The Monte-Carlo Method for Estimating the Uncertainty of the Surface Temperature Remote Retrieval

The Monte-Carlo method or the method of statistical modeling is a numerical solution of mathematical tasks with modeling random values. The wide practical application of the approach became possible with the use of contemporary fast computers.

The Monte-Carlo approach allows modeling practically any process that is influenced by random factors. However, these tasks do not restrict the field of its application. There are many other mathematical problems that seem not to be connected with rundom values and might be solved with inventing certain probabilitic models, with which their realization gives a needed solution. In some cases applying the artificial probabilitic simulations appears more effective (from the point of view of the algorithm simplicity and computer time expenditure) than a direct way of the problem solution. In the Chap. 13 the application of the Monte-Carlo approach will be considered for simulation for the process of radiation-atmosphere interaction and for calculation radiative characteristics in the atmosphere.

The Monte-Carlo approach is based on generating random numbers with prescribed statistical characteristics. It is possible to obtain a set of numbers imitating values of the random value η with a certain relation called a set of ***pseudorandom numbers***. If the set of not repeated pseudorundom numbers is long enough it is possible to assume these numbers as random.

8.3.1 Generating Pseudorandom Numbers

As mentioned above in order to solve mathematical problems using the Monte-Carlo method with the use of a computer, the algorithm for generating pseudorandom numbers offering prescribed statistical characteristics is used. *The probability density* $p(x)$ is the characteristic of a continuos random value η. *The distribution function* $F(x)$, *the mean of the random value* $M(\eta)$, and the dispersion $D(\eta)$ are derived from the probability density.

The probability density defines the probability of appearing the random value η within the interval (c, d):

$$P(c < \eta < d) = \int_c^d p(x)dx. \tag{8.16}$$

The probability density is to be positive within the definity ranges $a \le \eta \le b$, i.e. $p(x) > 0$, and it is valid for the whole interval:

$$\int_a^b p(x)dx = 1. \tag{8.17}$$

The mean of the random value is defined by the following expression:

$$\bar{\eta} = M\eta = \int_a^b x p(x)dx, \tag{8.18}$$

And *the dispersion* of the random value η is in accordance with the relation

$$D\eta \equiv M(\eta - M\eta)^2 = \int_a^b (x - \bar{\eta})^2 p(x)\, dx = \int_a^b x^2 p(x)dx - \bar{\eta}^2 \tag{8.19}$$

The distribution function of the random value η is the function:

$$F(x) = P(\eta < x) = \int_a^x p(x')\, dx'. \tag{8.20}$$

The random number defined at the interval (0, 1) and with the density distribution $p(x) = 1$ is especially important for the practical realization of the Monte-Carlo method. This random value is evenly distributed at the interval (0, 1) and will be denoted by the symbol γ. In this case $M(\gamma) = 1/2$, and $D(\gamma) = 1/12$.

There are a lot of algorithms in form of computer programs for generating pseudorandom numbers with even distribution at the interval (0, 1).

8.3.2 The Mathematical Simulation of the Influence of Initial Parameters Random Uncertainties on the Exactness of the Surface Temperature Retrieval $\mathbf{T_s}$

Calculating characteristics of the mathematical model reaction to randomly specified uncertainties of values regulating the model is demanded for studying the uncertainties' impact on the exactness of remote retrieving of the surface temperature.

The approach based on Eqs. 8.14 and 8.15 is used for estimating the uncertainty $|\Delta T_s|$, namely on the multiple modeling values $\Delta\alpha_0$ including in the Eq. 8.14 and obeing to normal distribution with the zeroth mean value and fixed dispersion (the mean square deviation is used in the computer program «Tsrnd» inspite the dispersion). Equations 8.7 and 8.8 are applied for calculating values T_s.

Statistical characterisitics of uncertainties ΔT_s, might be obtained in the program with modeling uncertainty for every parameter separately or any combination of parameters «Tsrnd». After specifying the needed set of parameters and their statistical characteristics 1,000 values of every parameter uncertainty are simulated with the standard subprogram *RND* (*random number digitizer*), and the uncertainty $|\Delta T_s|$ is calculated, then statistical characteristics of the $|\Delta T_s|$ are found. The following values are the output on the screen:

- hystogram of the uncertainties $|\Delta T_s|$;
- statistical characteristics of the $|\Delta T_s|$.

and in file:

- numbers of every uncertainty values for plotting the histogram;
- statistical characteristics of the $|\Delta T_s|$.

8.4 Practice 7

8.4.1 Objectives

1. To study theoretical basis of the remore retrieving of the surface temperature.
2. To study basis of mathematical simulation of physical processes described with the approach of statistical modeling and probability characteristics.
3. To obtain practical knowledge of applying the Monte-Carlo method for simulation of physical processes (case studies of initial data uncertainties' impact on the exactness of the remote surface temperature retrieval).
4. To compare application of two approaches ("deterministical" and "statistical" with the Monte-Carlo method) for estimation of the initial data uncertainty impact on the exactness of the remote surface temperature retrieval
5. To obtain numerical characteristics determining initial data uncertainties impact on the exactness of the remote surface temperature retrieval as a result of operating with computer numerical experiments.

Table 8.1 Versions of initial parameters

Version number	λ, µm	ε_v	P_v (p_s)	T_s, °C	$\tilde{T}_v$, °C
1	3.65	0.951	0.811	30	10
	8.1	0.951	0.811	30	10
2	3.66	0.822	0.912	28	9
	8.2	0.822	0.912	28	9
3	3.67	0.953	0.813	26	8
	8.3	0.953	0.813	26	8
4	3.68	0.824	0.914	24	7
	8.4	0.824	0.914	24	7
5	3.69	0.955	0.815	22	6
	8.5	0.955	0.815	22	6
6	3.7	0.826	0.916	20	4
	8.6	0.826	0.916	20	4
7	3.71	0.957	0.817	18	2
	8.7	0.957	0.817	18	2
8	3.72	0.828	0.918	16	4
	10.2	0.828	0.918	16	4
9	3.73	0.959	0.819	19	5
	10.3	0.959	0.819	19	5
10	3.74	0.961	0.92	12	1
	10.4	0.961	0.92	12	1
11	3.74	0.828	0.822	14	5
	10.4	0.828	0.822	14	5
12	3.74	0.961	0.822	16	8
	10.4	0.961	0.822	16	8
13	3.74	0.828	0.92	18	6
	10.4	0.828	0.92	18	6
14	3.74	0.961	0.822	19	7
	10.4	0.961	0.822	19	7
15	3.74	0.828	0.815	20	9
	10.4	0.828	0.815	20	9
16	3.75	0.822	0.822	21	9
	9.2	0.822	0.822	21	9
17	3.75	0.829	0.829	22	9
	9.2	0.829	0.829	22	9
18	3.75	0.827	0.825	23	9
	9.2	0.827	0.825	23	9
19	3.75	0.825	0.827	24	9
	9.2	0.825	0.827	24	9
20	3.75	0.826	0.825	25	9
	9.2	0.826	0.825	25	9

8.4.2 Software and Set of Input Parameters

1. Computer programs "V_VARTS.exe" and "VARTS.exe" for the first part of the practice, "V_TSRND.exe" and "TSRND.exe" for the second part.
2. The set of variants of initial parameters (the Table 8.1).

8.4.3 Test Questions

1. What initial parameters exactness does regulate the exactness of remote retrieving of the surface temperature?
2. What assumptions have been made for deriving the relation used for the calculation of the value T_s?
3. What initial parameter exactness does affect most dramatically on the exactness of remote retrieving the surface temperature?
4. Do uncertainties ΔT_s with the same value of the initial parameters uncertainties differ for different approaches? Why do they differ?
5. What defines the exactness of estimating statistical characteristics of the surface temperature T_s with the Monte-Carlo method?

8.4.4 Sequential Steps of the Exercise Implementation

1. To study the theory with using additional books, pointed in references list.
2. To choose a variant with a set of the initial parameters from the Table 8.1.
3. To create input data files with the program "V_VARTS.exe". Attention! The procedure is to be executed twice for two wave numbers. After completing the first data file "*vvarts.dat*" the calculation with the program "VARTS.exe" might be done. Then to repeat operation with program "V_VARTS.exe" and "VARTS.exe" for second wave number. The resulting output file "*varts.dat*" is written with the same name (in the second completing the output file is rewritten).
4. To prepare the plan of accomplishing numerical experiments with program "VARTS.exe" for

   - estimating the value $|\Delta T_s|$ with the use of two deterministic approaches;
   - analyzing the obtained results;
   - explaining the reason for the difference in values of $|\Delta T_s|$ arising with the use of the two approaches.

5. To accomplish the set of calculations with the program "VARTS.exe".
6. To plot and analyse the dependence of uncertainty of the surface temperature T_s remote retrieval on initial parameters uncertainties.
7. To study statistical characterisitics of uncertainty of the surface temperature T_s remote retrieval with the program "TSRND.exe" (it is to preliminary create the initial data file "*vtsrnd.dat*" with the program "V_TSRND") f.exeor different

values of mean square deviations of initial parameters assuming normal distribution of uncertainties of initial parameters and zeroth mean values, the result file is named "*tsrnd.dat*". Values of mean square deviations are specified within ranges 1–10% of parameters values. For specifying uncertainty of the average temperature $\tilde{T}_v$at the wave number v the value is taken in °K.

8.4.5 Requirements to the Report

Prepare the concise report with elements of theory, resulting pictures, and conclusions that reflect the main stages of the work. It is necessary to include in the report the following:

1. Presenting assumptions for the derivation of the Eq. 8.6.
2. Presenting the detailed derivation of the Eq. 8.11.
3. Explaining why the measuring of the intensity of outgoing radiation is to be accomplished at the transparency window.
4. Finding the initial parameters exactness providing the remote retrieving surface temperature uncertainty $|\Delta T_s|$ not larger than 0.5°K.

Chapter 9
The Thermal Remote Sounding of the Atmosphere

Abstract The inverse problem of the thermal sounding of the atmosphere is formulated. The matrix form of the problem is considered. Features of ill-posed inverse problem are analyzed. Two approaches for solution are proposed.

9.1 The Problem Statement

The thermal remote sounding of the atmosphere and surface is based on data of measurements of the heat radiance from boards of space platforms. And the problem arises: to retrieve the temperature vertical profile from observed data at the atmosphere top.

Thus ***the direct problem*** is solved at the first stage:

The solution of the direct problem is defined by the solution of the transfer Eq. 1.25 with ignoring scattering processes in heat spectral region ($\lambda > 3$ μm):

$$J_\nu^\uparrow(top,0) = \varepsilon_\nu B[\nu,T_s]P_\nu(top,0,\theta) + \int_0^{top} B[\nu,T(x)]dP_\nu(top,x,\theta) \tag{9.1}$$

where

$J_\nu^\uparrow(top,\ \theta)$ is the outgoing heat radiance at the atmosphere top (the level of satellite observation);

$\varepsilon_\nu B[\nu,T_s]P_\nu(top,0,\theta)$ is the heat radiation emitted by the surface with the temperature T_S and decayed by the atmosphere (the surface yield to the heat outgoing radiation);

$\int_0^{top} B[\nu,T(x)]dP_\theta(top,x,\theta)$ is the heat radiance of all layers of the atmosphere with the temperature $T(x)$, decayed by above layers (the atmosphere yield to the heat outgoing radiation);

I. Melnikova et al., *Remote Sensing of the Environment and Radiation Transfer*,
DOI 10.1007/978-3-642-14899-6_9,

It is necessary to specify following parameters: ε_ν, T_s, P_ν, dP_ν for the calculation the outgoing radiance $J^\uparrow$ at the wavelength $\lambda[\mu m] = 10{,}000/\nu[cm^{-1}]$ where:

ε_ν is the surface emissivity, its values are in range [0,1],
T_s is the surface temperature,
P_ν (a,b,θ) is the transmission function of the atmospheric layer $[a,b]$,
dP_ν is the differential of the transmission function,

The differential of the transmission function demonstrates the velocity of its variation at the level x, namely $P_\nu(x, top, \theta) - P_\nu(x + dx, top, \theta)$,

θ is the viewing angle.

Let the surface temperature T_s and the surface emissivity ε_ν are known, the transmission function $P_\nu(a,b,\theta)$ and its differential dP_ν depend on the absorption coefficient of the atmosphere and might be calculated using the models of gaseous absorption bands and known content of the corresponded gas: ρ_{gas}/ρ_{air} is the gas specific content. It is possible to assume that carbon dioxide (CO_2) content is constant over altitude and the transmission function $P_\nu(a,b,\theta)$ is close to 1 (unit) within transparency window.

9.2 The Analysis of the Direct Problem

Different over altitude layers in the atmosphere give dissimilar contribution to forming the outgoing radiation in various spectral channels. Consider physical precondition of temperature profile retrieval taking in mind the absorption bands of carbon dioxide CO_2.

Let the channel ν_1 correspond to the strong gaseous absorption and the radiance $J^\uparrow{}_1$ is formed,

the channel ν_2 correspond to intermediate absorption with the radiance $J^\uparrow{}_2$,
the channel ν_3 correspond to the weak absorption with the radiance $J^\uparrow{}_3$

In case of ***the strong absorption*** the transmission function differential is not equal to zero only in the top layer $z_i - z_n$, and in other layers P_ν and dP_ν are equal to zero.

In spectral channels ν_2 with ***the medium absorption*** the transmission function differential is not equal to zero only in the middle layer, and at channels with ***the weak absorption*** ν_3 the differential is not equal to zero dP_ν close to the surface.

Thus it is possible to solve *the inverse problem* from measuring the outgoing heat radiance using the Eq. 9.1. The inverse problem solution might give the temperature profile $T(x)$ in the atmosphere.

9.3 Possibilities for the Inverse Problem Solution

Above the inverse problem for the surface temperature retrieval was solved and one value T_s was obtained. Now the desired solution is a function, generally speaking infinite set of temperature values over the altitude. The temperature vertical dependence $T(x)$ is not an analytical function, however it is possible to specify with a finite set of values (about 40–50 values) at certain altitudinal levels. At some geographical sites there are data of aerologic soundings. It is clear that it is impossible to obtain 40 values of the function from one value (observational data). Thus it is necessary to accomplish 40 observations at different spectral channels. Assume for the solution the following:

1. Let's consider that the surface temperature T_s is obtained from observations at the quasi-transparent spectral regions that lie between the major line clusters and called *windows*;
2. There is a geographical bridging of satellite observational site and the surface emissivity ε_ν is known for the specific surface;
3. It is necessary to know the transmission function $P(x)$, it is possible if the radiation is measured at CO_2 bands because its concentration does not vary and CO_2 profile is known. CO_2 bands correspond to spectral intervals 2, 3.7, 4.3, 6.3, 8, 12 and 15 μm. It is to choose wave numbers in such a way as to embrace three spectral diapasons with strong, medium, and weak absorption for obtaining the temperature $T(x)$ at different altitudes.

The unknown function $T(x)$ is included in the Planck's function under the integral, thus it is demanded to factor the Planck's function outside the integral sign. It is more effective to find the deviation of the temperature from the average value at every altitudinal level x.

$$\Delta T(x) = T(x) - \bar{T}(x), \tag{9.2}$$

where $\bar{T}(x)$ is the average temperature profile obtained from long-standing aerologic observations.

After writing the Eq. 9.1 for the average temperature profile $\bar{T}(x)$ and considering the difference $\Delta T(x)$ functions might be expanded into a Taylor series over small values $\Delta T(x)$. Then the equation might be derived after keeping only the first item in the expansion and neglecting items with higher power.

$$\Delta J_\nu{}^\uparrow = J_\nu{}^\uparrow - \bar{J}_\nu{}^\uparrow \cong \int\limits_0^{z_\Pi} \frac{\partial B}{\partial T}\bigg|_{T=\bar{T}(x)} \Delta T(x) \frac{\partial P_\nu(x, top, \theta)}{\partial x} dx, \tag{9.3}$$

Let's analyze the Eq. 9.3:

The left part of the Eq. 9.3 contains the values $J^\uparrow{}_\nu$ that is the result of observation and $\bar{J}^\uparrow{}_\nu$ is the result of calculation for the average temperature profile.

The right part of the equation contains functions depending on altitude x (vertical profiles):

$\Delta T(x)$ is the function of real temperature profile deviation from the calculated one.

$\frac{\partial B}{\partial T}\Big\|_{T=\bar{T}(x)}$ is the value of the Planck's function differential over the temperature at the point $\bar{T}(x)$ that is possible to calculate at all levels x, with knowing the average temperature values (average profile);

$\frac{\partial P_v(x,top,\theta)}{\partial x}$ is the known function because the CO_2 content and absorption coefficient at different spectral intervals are noted;

The Eq. 9.3 is the integral Fredholm equation of the first kind:

$$f(x) = \int_a^b K(x,y)\phi(y)dy \qquad (9.4)$$

where $f(x)$ is the known function, $K(x)$ is the kernel of integral equation, $\varphi(x)$ is the desired function. In our case noting corresponds to the following functions:

$$f(x) = \Delta J_v^{\uparrow}, \quad \phi(y) = \Delta T(x), \quad K(x,y) = \frac{\partial B}{\partial T}\Big\|_{T(x)=\bar{T}(x)} \frac{\partial P_v}{\partial x}$$

The function $\frac{\partial B}{\partial T}\Big\|_{T=\bar{T}(x)}$ is easy to prescribe analytically, however the function $\frac{\partial P_v(x,top,\theta)}{\partial x}$ does not allow an analytical presentation. Thus the additional transformations are needed.

Exchange the integral item by its finite-dimensional analogue – the sum. The integrating interval (the atmosphere over altitude) is split into N levels, the temperature T and transmission function P are specified at chosen levels.

$$\int_a^b l(x)dx \approx \sum_{i=0}^{N} l_i \Delta x_i w_i,$$

where w_i is the weight function, which depends on the numerical integrating approach (methods of "rectangulars", "trapeziums" or "polynomials").

It is necessary to specify transmission function values $P_{v\ i}$ at every level x_i and calculate corresponding differences

$$\Delta P_v(x_i) = P_v(x_i) - P_v(x_i) \quad \Delta T(x_i) = T(x_i) - \overline{T}(x_i)$$

Then the Eq. 9.3 is presented as follows:

$$\Delta J_v{}^{\uparrow} = \sum_0^N \Delta T_i \frac{\partial B}{\partial T}\Big\|_{T=\bar{T}(x)} \Delta P_v \qquad (9.5)$$

The equation might be written for 50 levels in the atmosphere:

$$\Delta J_\nu{\uparrow} = \sum_{0}^{N} \Delta T_i K(\nu, x_i), \quad N = 50 \tag{9.6}$$

It is apparent that many observations are required at different wave numbers: $J^{\uparrow}_{\nu_i}$, $i = 1,2,3,\ldots$ and in different spectral intervals for creating the system of linear algebraic equations. Wave numbers ν_i are to be chosen at transparency windows and at intervals of weak, medium, and strong absorption for retrieving the surface temperature and the temperature at different altitudes.

9.4 The Matrix Form of the Inverse Problem

Lets introduce the following notes:

$\vec{f} = \left\{\Delta J^{\uparrow}_{\nu i}\right\}$ is the vector of deviations observed radiance from the radiance calculated for average temperature profile at corresponding wave numbers, $m \sim 15$

$\vec{\phi} = \{\Delta T_i\}$ is the vector of temperature deviations from average values at corresponding altitudinal levels in the atmosphere, $n \sim 50$

Introduce the matrix A, which elements presents values of the Planck's function differential over temperature at corresponding altitudes and values of the transmission function at corresponding spectral intervals (defining the absorption coefficient values $k_{\nu j}$) and at altitudinal levels (defining CO_2 content):

$$A[m \times xn]; \left|a_{i,j}\right| = K(\nu_j, z_i)$$

Then the equation system (9.6) for the array of m observations is written in the matrix form:

$$\vec{f} = A\vec{\phi}, \tag{9.7}$$

where:

$\vec{\phi}$ is the desired vector at level i with dimension n, (temperature deviations at n altitude levels)

$\vec{f}$ is the vector of deviations between observed and the radiance calculated for averaged temperature profile of order m (in m spectral intervals),

A is the matrix of order [m × n], with elements $K(\nu_j, z_i)$.

Reminder basing on the definition, rules, and relations from *the linear algebra*

A matrix is called *rectangular* if m (number of rows) is not equal to n (number of columns).

A matrix is called *square* if $m = n$

In the *transposed form* of a matrix B is denoted by B*, the element $(B)_{ji}$ in the j'th row and i'th column of B is equal to the element $(B^*)_{ij}$ in the *i*'th row and *j*'th column of B*. Formally $(B^*)_{ij} = (B)_{ji}$

$$\underset{[m \times n]}{B} \rightarrow \underset{[n \times m]}{B^*}$$

The inverse matrix B^{-1} for the matrix B (it is unique only for square matrix) if the equality $(B)_{ij}(B)_{ij}^{-1} = 1$ is valid for elements of the inverse matrix.

The unit matrix is the result of multiplying square matrix to the inverse matrix:

$$B \times B^{-1} = E \quad \begin{vmatrix} 1 & 0 & 0 \\ 0 & 1 & 0 \\ 0 & 0 & 1 \end{vmatrix} \quad [m \times m][m \times m]$$

Two matrices C and D can be *multiplied* if the number of columns in C equals the number of rows in D. Let C be of order $[m \times l]$ (have *m* rows and *l* columns) and D of order $[l \times m]$. Then the product of two matrices K = CD, is a matrix of order $[m \times m]$

$$\underset{[m \times l]}{C} \times \underset{[l \times m]}{D} = \underset{[m \times m]}{K}$$

Multiplication of a matrix to a *vector*:

$$\underset{[m \times 1]}{\vec{f}} = \underset{[m \times n]}{A} \times \underset{[n \times 1]}{\vec{\phi}}$$

Let's consider the auxiliary equation for further transformations:

$$\underset{[m \times 1]}{\vec{a}} = \underset{[m \times m]}{C} \underset{[m \times 1]}{\vec{d}}$$

Multiplication of the equation by the inverse matrix C^{-1} leads to the result:

$$C^{-1}\vec{a} = C^{-1}\, C \quad \vec{d} = \vec{d}$$

because from definition $C^{-1} \times C = E = \begin{bmatrix} 1 & & 0 \\ & 1 & \\ 0 & & 1 \end{bmatrix}$ it is valid: $\begin{bmatrix} 1 & & 0 \\ & 1 & \\ 0 & & 1 \end{bmatrix} \begin{matrix} d_1 \\ \cdots \\ d_m \end{matrix} = \begin{matrix} d_1 \\ \cdots \\ d_m \end{matrix}$ with taking into account rules of multiplying matrix to vector.

The result of solution of the equation for square matrix is $\vec{d} = C^{-1}\vec{a}$.

Thus it is required to find an inverse matrix for the matrix A for solving the equation. Point out that the diagonal matrix is easy transforming to the inverse one according to the rule:

$$\begin{vmatrix} a_1 & & & 0 \\ & a_2 & & \\ & & a_3 & \\ 0 & & & a_4 \end{vmatrix}^{-1} = \begin{vmatrix} \frac{1}{a_1} & & & 0 \\ & \frac{1}{a_2} & & \\ & & \frac{1}{a_3} & \\ 0 & & & \frac{1}{a_4} \end{vmatrix}$$

Let's multiply the Eq. 9.7 by the transposed matrix A^* for obtaining the square matrix in spite of matrix A

$$A * \times \vec{f} = A * \times A \times \vec{\phi}$$

dimensions: $[n \times m]$ $[m \times 1]$ $[n \times m]$ $[m \times n]$ $[n \times 1]$

Then multiply the equation to the square matrix $(A^* A)^{-1}$

$$(A * A)^{-1} A * \vec{f} = (A * A)^{-1} (A * A) \vec{\phi} \tag{9.8}$$

The combination $(A * A)^{-1}(A * A) = E$ is equal to the unit matrix, and the solution of the Eq. 9.8 looks as follows:

$$\vec{\phi} = (A * A)^{-1} A * \vec{f} \tag{9.9}$$

dimensions

$$[n \times 1] \quad \underbrace{[m \times m] \quad [n \times m]}_{[m \times n]} \quad [m \times 1]$$

Let's test the dimensions of the left and right equation parts. The following cases might be:

1) $m = n$; 2) $m > n$; 3) $m < n$.

In the simplest case $m = n$ the system of linear equations is defined by square matrix and there exists a unique solution (9.9). Nevertheless this solution appears inappropriate because the Eq. 9.7 characterizes an *ill-posed problem.*

The ill-posed problem has appeared when the arbitrary infinitesimal variation of initial data provokes arbitrary high variation (uncertainties) of the solution.

Earlier they supposed that it is just not correctly formulated problem; however in the middle of the last century it was determined that there is a special big enough class of ill-posed problems. For example problems of interpretation seismic data and problems of the remote sensing (thermal tomography, aerosol retrieval from optical and lidar data) are ill-posed problems. Andrey Tikhonov known Soviet mathematician published the article in 1943 where *the theory of ill-posed inverse*

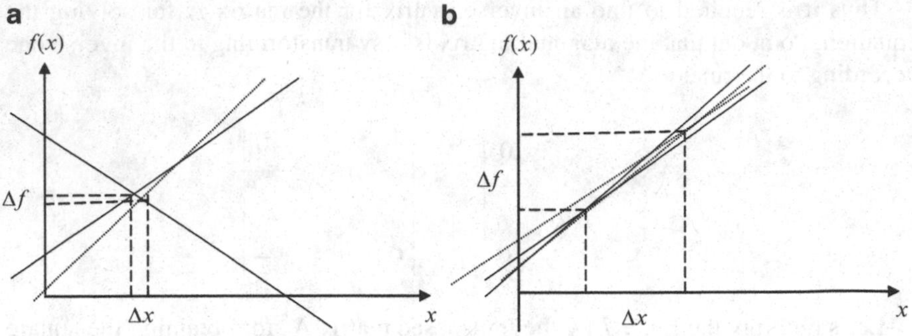

Fig. 9.1 The solution of the system of two linear equations in Cartesian plane: (**a**) the small deviations of the curves intersecting at a big angle does not lead to significant deviation Δf and Δx; (**b**) small deviations of curves intersecting at small angle calls big deviations Δf and Δx

problems solution has been formulated. Figure 9.1 demonstrates the solution of system of two linear equations in Cartesian plane. In the Fig 9.1a the small deviations of the curves intersecting at a big angle does not lead to significant deviation Δf and Δx. In the Fig. 9.1b small deviations of curves intersecting at small angle calls big deviations Δf and Δx.

In our case the function ϕ (the vector ϕ is a set of values for the set of altitudes) is a solution of the Eq. 9.9. The direct use of this solution gives the picture presented in the Fig. 9.2. The solid line demonstrates the direct solution (9.9) of the Eq. 9.7.

Variations of desired values might be arbitrary big. The dashed line is the real values of the temperature at corresponding altitude levels. Both vectors $\hat{\vec{\phi}}$ and $\vec{\phi}$ give the same vector $\vec{f}$, when substituted in the Eq. 9.7 because of averaging called by relation of matrix and vector dimensions that provokes an ambiguity of solution.

Let us consider two approaches for the inverse ill-posed problem solution:

The method of maximum smoothness
The method of statistical regularization

9.5 Solution of the Ill-Posed Inverse Problem of the Remote Temperature Sensing of the Atmosphere

The remote temperature sensing is the obtaining the temperature profile in the atmosphere. The corresponding inverse problem arises $\vec{f} = A\vec{\phi}$. The formal solution (9.9) appears invalid thus the additional information is needed. Let us analyze the vector $\vec{\phi}$. It defines the temperature profile and we have to use known a priori statistical properties of the desired vector for improving the solution.

Over many years averaged values are possible to calculate for every season and the temperature correlation matrices could be constructed.

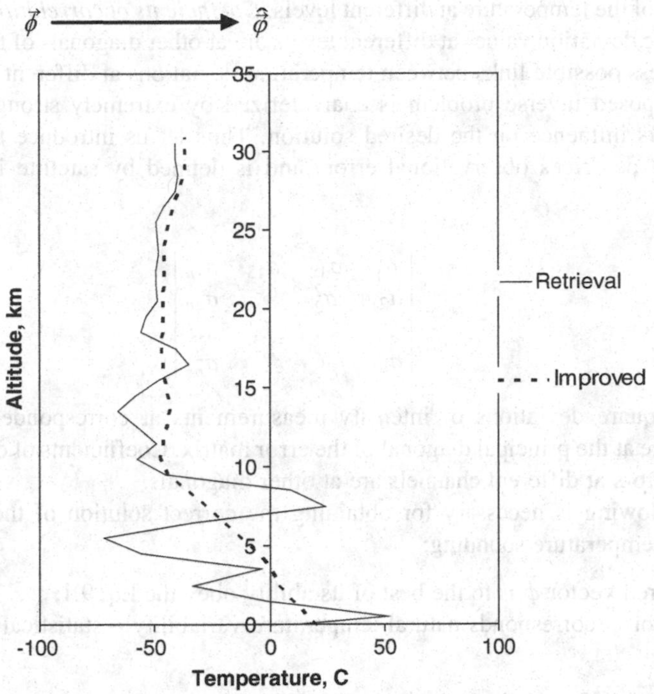

Fig. 9.2 The solution of the ill-posed inverse problem of remote sensing. Strict formal solution – *solid line* and solution corrected with the temperature correlation matrix – *dotted line*

Let the temperature profile be obtained at North-West of Russia. Then the data base of aerologic sounding at the Main Geophysical Observatory in Voeykovo, St. Petersburg suburb that regularly widened for many years, provides calculating values of the temperature at different altitudes averaged over many years in corresponding season and constructing the temperature correlation matrix K_{TT}. The Fig. 9.2 demonstrates vertical profiles of the desired vector $\vec{\phi}$ (solid line) and values $\vec{\hat{\phi}}K_{TT}$ corrected with taking into account for the temperature correlation matrix (dashed line). The mean temperature profile is obtained with averaging over e.g. 50 observed values at every altitude in corresponding season and interpolating at needed levels. The correlating matrix of temperature deviations from the mean value at every level is constructed as follows:

$$K_{TT} = \begin{vmatrix} s_1^2 & k_{12} & k_{13} & k_{1n} \\ k_{21} & s_2^2 & & \\ k_{31} & & s_3^2 & \\ k_{n1} & & & s_n^2 \end{vmatrix}$$

Averaged real temperature deviations from mean temperature values at every altitude are at the principal diagonal of the square matrix. They express a natural variability of the temperature at different levels. *Coefficients of correlation* between temperature deviation values at different levels are at other diagonals of the matrix. They express possible links between temperature deviations at different altitudes.

The ill-posed inverse problem is characterized by extremely strong observational errors influence on the desired solution. Thus let us introduce the error's matrix that describes observational errors and is defined by satellite instrument quality.

$$\Sigma_{TT} = \begin{vmatrix} \sigma_1^2 & \sigma_{12} & \sigma_{13} & \sigma_{1n} \\ \sigma_{21} & \sigma_2^2 & & \sigma_{2n} \\ \dots & \dots & \dots & \dots \\ \sigma_{n1} & & & \sigma_{nn}^2 \end{vmatrix}$$

Mean square deviations of intensity measurements at corresponded spectral channels are at the principal diagonal of the error matrix. Coefficients of correlation between errors at different channels are at other diagonals.

The following is necessary for obtaining the correct solution of the problem of remote temperature sounding:

1. the desired vector $\vec{\phi}$ is to the best of its ability obey the Eq. 9.1;
2. the vector $\vec{\phi}$ corresponds natural temperature variability – statistical ensemble K_{TT}.

The method **of *statistical regularization*** is used to satisfy these two conditions. Then *the regularized solution* takes the form:

$$\hat{\vec{\phi}} = \left(A^T\Sigma^{-1}A + K_{TT}^{-1}\right)^{-1} A^T\Sigma^{-1}\vec{f} \tag{9.10}$$

This solution (9.10) is the result of an entire mathematical course: methods of solving ill-posed problems.

It is necessary to point out that the vector $\hat{\vec{\phi}}$ is not rigorous and exact solution of the Eq. 9.1 and when substituting to the equation, it provides only approximate equality. But it is clear that the equality basically could not be rigorous because of observational errors.

$$\vec{f} \approx A\hat{\vec{\phi}} \tag{9.11}$$

However, values of the vector $\hat{\vec{\phi}}$ are physically justified though do not provide the rigorous equality (9.1). The abundance of a priori temperature profiles (rich statistics) is required for the physical justification.

It's not always possible to have rich statistics, e.g. over oceans, in atmospheres of other planets. Then the *method of maximum smoothness* is applied.

It is a priori known that temperature (and any other physical value) profile in the atmosphere have to be smooth and escapes leaps because atmospheric properties vary smoothly. The property of smoothness may be formulated as temperature values at two neighbor levels weakly differ. The mathematical formulation of smoothness equals to small derivatives of the temperature with respect to altitude. Thus *the method of maximum smoothness or Tikhonov method* is applied.

The matrix H which is analogues to numerical differentiation when it's multiplied by the vector. It is assumed as follows:

$$\mathrm{H}\vec{\phi} = \begin{vmatrix} 1 & -1 & 0 & 0 \\ & 1 & -1 & 0 \\ & & 1 & -1 \\ & & -1 & 1 \end{vmatrix} \times \begin{vmatrix} \phi_1 \\ \phi_2 \\ \\ \phi_n \end{vmatrix} = \begin{vmatrix} \phi_1 - \phi_2 \\ \\ \\ \phi_{n-1} - \phi_n \end{vmatrix},$$

The resulting vector characterizes the profile smoothness at neighboring altitudinal levels. Differences ϕ_1–ϕ_2 are to be minimal for the best result. The solution could be presented as:

$$\hat{\vec{\phi}} = \left(\mathrm{A}^*\Sigma^{-1}\mathrm{A} + \alpha\mathrm{H}\right)^{-1}\mathrm{A}^*\Sigma^{-1}\vec{f} \tag{9.12}$$

where the value α is the regularization parameter according to Tikhonov. By changing the parameter α the weight of the average profile and observational data is balanced. The parameter α is not known a priori and requires to be found. The approach of *numerical closed successive experiment* is used for α defining.

9.6 Numerical Closed Successive Experiment

Let the temperature deviation profile (vector $\vec{\phi}$) be known. It is easy to calculate the temperature profile $\vec{T}(x)$ and then to obtain the outgoing intensity with the Eq. 5.1: $\vec{T}(x) \to J^{\uparrow}$, i.e. *the direct problem* is solved. With the averaging temperature profiles over the data base the mean temperature profile is obtained and the corresponding intensity is calculated: $\vec{\bar{t}} \to \bar{J}^{\uparrow}$, and then the difference $\Delta J = J^{\uparrow} - \bar{J}^{\uparrow}$, that is equal to the vector $\vec{f}$. The set of calculated values $\vec{f}$ ("simulated observations") and then retrieved vectors $\vec{\phi}(\alpha_1)$ and $\vec{\phi}(\alpha_k)$ are accomplished for the set of parameters α_k. The optimal value of the regularization parameter α is derived with scanning a lot of values α (e.g. 500), then different temperature profiles (1,000) are scanned and 1,000 optimal parameters α are obtained and at last the mean optimal value α is calculated with averaging over all realization for the needed season and region, which is used for solving the remote sensing problem.

In practice aerologic sounding data (e.g. 500–600 profiles) are used for deriving optimal α from every profile, then calculating the average value.

9.7 Practice 8

9.7.1 *Objectives*

To study specific features of methods applied for regularizing the solution of the inverse ill-posed problem of remote temperature sounding.

To obtain the optimal solution by choosing input parameters.

9.7.2 *Software and Set of Input Parameters*

1. **The method of *statistical regularization*** supposes analyzing matrices Σ и K_{TT} influence on the desired result. There are two computer programs in the directory Lab4:

   The program "RADNIMB6.exe" solves the direct problem, and calculates the outgoing intensity from input temperature profile "aer.dat" that contains transmission function and average temperature profile $T(z)$.

   The program "DZ_T_SR5.exe", solves the inverse problem for choused matrices Σ and K_{TT} elements. All elements at principal diagonal are taken equal each other (only one value is taken) and correlation coefficients are zeroth.

   Variants of input data

   There are two temperature profiles: mean and aerologic (true) T_{aer}. The transmission function P is calculated at corresponding spectral channels. For example:

ν_{ij}, sm^{-1}	P
669.8	0.001 strong absorption
676.7	0.8 weak absorption
…	…
746.7	0.22

   Four variants of input parameters are to be considered:

   (a) **variant:** All diagonal elements of the temperature correlation matrix K_{TT} are equal to 0.1

   Values of diagonal elements of the errors matrix Σ are equal to 1.0

   The program calculates the vector $\hat{\vec{\phi}}$ (retrieved temperature deviations profile), then calculates temperature values $\hat{T}$, and at last the difference $\hat{T} - T_{aer}$. Result is saved in the file "*dz.dat*" In the first variant the solution coincides with the mean profile. The mean deviation from true values appears around 8°C because there is no new information.

   (b) **variant**: diagonal elements of the temperature correlation matrix K_{TT} are equal to 10 (i.e. average deviations from the mean profile are ±3°)

   Values of diagonal elements of the errors matrix Σ are equal to 0.1.

The result is closer to the true profile: average deviation is about 3°C.

(c) **variant:** Diagonal elements of the temperature correlation matrix K_{TT} are equal to 10, i.e. result could considerably differ from the mean profile. Values of diagonal elements of the errors matrix Σ are equal to 10 that points big observational errors. The result coincides with the mean profile because observational data are suspicious.

(d) **variant:** Allowing that desired result strongly differ from mean profile ($K_{TT} = 20$) and assuming that the high exactness of observations ($\Sigma = 0.0000001$) we obtain the result close to the formal, rigorous solution (9.6) and unreliable physically impossible variations of the temperature (±3,000 K) at different altitudes demonstrates main feature of the ill-posed problem.

One should inquire into the exactness of the instrument, which is put on a satellite board for adequate specifying the matrix Σ. The instrument on a satellite board degrades and the exactness deteriorates. Thus the matrix Σ is to be corrected while satellite is on the orbit.

Obtained here results demonstrate a strong deviation from the real temperature profile at the tropopause in all cases of initial parameters, where the temperature varies dramatically. In our case only one value is assumed for all altitudes in spite of the set of values defining the matrices K_{TT} or Σ. For better temperature retrieval at the tropopause more spectral channels are provided. The real exactness of the temperature remote retrieval is reached at about 0.5–0.7°C. It is important to choose a successful alternative between variants: smoothness of profile and closeness to the mean profile and very strong intrusion of the mean profile and loss of the real information. It is clear that a rich volume of the a priori information is necessary (a lot of correlation matrix for any season and any geographical site) for a reliable temperature retrieval.

2. **The method of maximum smoothness.**
The program "RADNIMBUS.exe" (in directory Lab4) accomplishes the auxiliary calculation for the taken temperature profile. Then the program "DZ_T_FU1.exe" solves the inverse problem using the method of maximum smoothness. Result is saved in the file "*fu.dat*". The ranges of the regularization coefficient α variation is proposed as 0.00001–0.1. Values of all elements of the matrix Σ are $\sigma_{ij} = 1$ corresponds to rough observation with significant errors; $\sigma_{ij} = 0.1$, corresponds to more exact observation.

The result of the program operation leads to the optimal value of the regularization coefficient $\alpha_{\text{optimal}} = 0.005$, and mean square deviation 5.1°. The next result presents three profiles: true, mean and retrieved. The error appears 8° and it is too big.

The following iteration is close to the regularization coefficient value $\alpha = 0.005$ with the range of variation 0.001–0.01. Values of matrix Σ elements $\sigma_{ij} = 0.1$ give the optimal value of the regularization coefficient: $\alpha_{\text{optimal}} = 0.001$; and the mean square deviation 3.1°. Thus this retrieved profile is better compatible with the true one.

The following iteration uses ranges of the regularization coefficient 0.0005–0.0015, and values of matrix Σ elements $\sigma_{ij} = 0.1$, that gives $\alpha_{\text{optimal}} = 0.0006$, the mean square deviation 2.95°, that improve the result by 0.15°C.

The results on the display show that retrieved temperature profiles significantly differ from the true one because we use only one value 0.1 instead of the correlation matrix.

The method of maximum smoothness is recommended in case of a lack of the a priori detailed information about temperature profiles (e.g. in the atmospheres of other planets). *The method of statistical regularization* is more effective in the Earth's atmosphere where the rich a priori statistics is available.

The result after considering both method for the same profile and equal initial parameters is the following:

The method of maximum smoothness gives $\alpha_{\text{optimal}} = 0.005$, and the mean square deviation 5.8°;

The method of statistical regularization provides the mean square deviation 5.47°.

It is seen that the method of statistical regularization provides somewhat better exactness of temperature profile retrieval. Both methods give physically proved solutions.

9.7.3 Sequential Steps of the Exercise Implementation

1. Prepare the table with the atmosphere aerologic sounding data.
2. Use file in the database with name "*aer.dat*". To plot the initial temperature profile in EXCEL.
3. To calculate the outgoing intensity with the program "RADNIMB.exe" in six spectral channels corresponding to spectral intervals of the radiometer «Nimbus». The following files in the directory Lab4 are needed for operating program "RADNIMB.exe":
   - "*nimbus. dat*";
   - the file with aerological data: "*aer.dat*".

The "RADNIMB" program operating create the file "*nim.dat*" in the directory Lab4, which is necessary for operating programs "DZ_T_SR.exe" and "DZ_T_FU.exe". It is possible to use ready files "*nim.dat*".

To plot profiles of the transmission function differential dP and the product dP * B in EXCEL-editor.

To create the table of outgoing intensity for mean and true (aerological) temperature profiles, calculated with "RADNIMB.exe" program.

1. To prepare the auxiliary report with obtained results.
2. To retrieve the temperature profile with the program "DZ_T_SR.exe" using the method of statistical regularization from "observations" (calculation with

program "RADNIMB.exe") of outgoing heat radiation for a set of matrices Σ и K_{TT} elements. To plot obtained profiles.

3. To obtain the temperature profile with the program "DZ_T_FU1.exe" using the method of maximum smoothness. To study the influence of the regularization coefficient α and diagonal element of the matrix Σ on the exactness of the temperature retrieval. To plot retrieved profile and compare it with the true and mean ones.
4. To compare results of considered methods.
5. To prepare the final report.

Chapter 10
Calculating Optical Characteristics of Atmospheric Aerosol

Abstract Calculations of optical characteristics: scattering and absorption coefficients and phase function of modelled ensembles of aerosol particles are considered. The description of the practice is given.

10.1 Atmospheric Aerosol

An equilibrium mixture of gas and solid or liquid particles in the atmosphere is called atmospheric aerosol (sometimes they use plural aerosols for particle variability). Atmospheric science implies particles itself in the air. Sources of atmospheric aerosol are extremely varied: dust lifted into the atmosphere by wind (mineral, silicates), sea-salt crystals remaining after sea water droplets evaporates, products of chemical reactions in the atmosphere (including photochemical reactions producing smog), volcanic eruptions, product of fires (soot aerosols) and anthropogenic (industrial) pollution. According to the definition clouds and fogs are aerosols too, but usually they are separated to a special class of atmospheric objects.

Atmospheric aerosol plays an important role in radiative processes in the atmosphere. It interacts (absorbs and scatters) solar and heat radiation and determines atmospheric optical properties. This interaction is especially essential in the shortwave spectral ranges (from UV till near infrared), right where the maximum of the solar flux is situated. There is an important feature of the aerosol – its strong space and temporal variability: the aerosol content in the air of big cities is two-three orders higher than in the clean air, heavy volcanic eruptions exhaust a lot of matter to the middle atmosphere and change optical properties of the stratosphere for months and even years. Thus the stipulated importance of studying and numerical simulating of aerosol optical properties is undoubted.

I. Melnikova et al., *Remote Sensing of the Environment and Radiation Transfer*, DOI 10.1007/978-3-642-14899-6_10,

10.2 Interaction Between Radiation and Aerosol Particle

Separate aerosol particle is to be modeled as an object of definite shape for mathematical description. The simplest shape is a sphere.

The problem of calculating electromagnetic waves interaction with homogeneous sphere was solved in 1908 by German physicist Gustav Mie, and derived a theory called *the Mie theory*. According to the Mie theory all necessary optical characteristics of the spherical particle are obtained by assuming that the relation $y = 2\pi r/\lambda$, where r is the radius of the sphere and λ is the wavelength of incident radiation, as well as the *complex refractive index* (CRI) of the sphere matter $y = 2\pi r/\lambda$. The meaning of m is considered in details in the book. Here we'll only clarify that the real part of CRI is the refractive index (the ratio of light velocity in a vacuum and in the matter), and imaginary part characterizes the radiation absorption by the particle matter.

Resulting formulas of the Mie theory are cumbersome and are not presented here, details are in books. For 100 years a significant additional work has been done for transforming Mie formulas to convenient forms for calculations. Finally we use the algorithm for computer codes.

Hence, input data are: the parameter $y = 2\pi r/\lambda$, and CRI of the particle matter m (λ) depends also on wavelength λ. Output data are assumed the cross-section of interaction particle and radiation: the extinction cross-section $C_e(y,m)$; the scattering cross-section $C_s(y,m)$; absorption cross-section $C_a(y,m)$ and the scattering phase function $x(\gamma,y,m)$, where γ is the scattering angle. It is to be mention that the aerosol absorption is not selective as distinct from gas absorption. The rigorous definition of interaction cross-sections is in references (Joseph et al. 1976). Here we'll only clarify the physical meaning of the notion.

10.3 Ensemble of Aerosol Particles

While applying the Mie theory to problems of atmospheric optics it is to be accounted for that real aerosol particles range in size from about 10^{-4} μm to tens of micrometers. This property is called *the particle dispersivity*. The collection of particles of all possible dimensions is called *the ensemble of particles*.

The characteristic of aerosol particles number in the air is the concentration: the number of particles in the volume unites. Depending on the particle size and the geographical location particle concentration ranges from about 10^7 to 10^{-6} cm^{-3}. The aerosol dispersivity leads to particle concentration being inadequate for the ensemble describing. It is evident that particles of different sizes have different concentrations. Hence it is yet one characteristic to be introduced for linking the particle concentrations and radius.

Let the number of all particles (in the volume unite) with radius less or equal r be $N(r)$. Then particle number with radiuses in range from r to $r + \Delta r$ is

$N(r + \Delta r) - N(r)$. Let's turn to mean concentration i.e. to the value $[N(r + \Delta r) - N(r)]/\Delta r$ for ability to compare particle concentration in different intervals Δr. Let Δr decrease towards zero and obtain, according to the derivative definition, *the function of particle size distribution* $n(r) = dN(r)/dr$. The distribution function $n(r)$ completely describes the aerosol particle ensemble considering its dispersivity and allows finding the concentration of any radius range (in particular the total aerosols concentration)

$$N = \int_0^\infty n(r)dr \tag{10.1}$$

The function $f(r) = n(r)/N$ is *the normalizing function of aerosol particles distribution*, which is convenient because allows considering particle dispersivity independently of specific concentration N. Often it is called just *the distribution function* without the refinement *normalizing*. We will use this terminology in further consideration. *The normalizing condition* for the function $f(r)$ with the Eq. 10.1 is the following

$$\int_0^\infty f(r)dr = 1 \tag{10.2}$$

The function $f(r)$ offers also a sense of probability: it is *the probability density* of aerosol particle has the radius r.

The experimental data of aerosol concentration versus particle radius $N(r)$ are used for describing size distribution function of real atmospheric particles ensembles. For convenience the experimentally obtained distribution function $f(r)$ is approximated with analytical formulas. The normal logarithmic (lognormal) distribution is the mostly used approximation:

$$f(r) = \frac{1}{\sqrt{2\pi}rs}\exp\left(-\frac{1}{2s^2}\ln^2(r/r_0)\right) \tag{10.3}$$

Lognormal distribution (10.3) is the distribution of the value that logarithm changes according to normal (Gaussian) law. It is fully characterized with two parameters: *the mean radius* r_0 (as a logarithm value) and *the dispersion* s. The mean radius describes particle sizes in total (the larger the radius r_0) the greater are in average particles of the ensemble. The dispersion s represents the scatter of radiuses (the larger the parameter s, the greater the difference between concentrations of particles with various radiuses). We use the lognormal distribution (10.3) in the consideration.

10.4 Calculation of Optical Characteristics of Aerosol Particles Ensembles

For monodispersal distribution (all particles are the same size) the simple relation is used to transit from one particle characteristic (e.g. the extinction cross-section C_{e},) to the ensemble characteristic (the volume extinction coefficient α)

$$\alpha = NC_e, \tag{10.4}$$

where N is the particle concentration (the number in the volume unite). The relation (10.4) expresses physically *the summation principle*, i.e. cross-section of individual particles are summed (as shadow squares) to the total characteristic of extinction of the volume unite (summed shadow of all particles). The summation principle provides the extension of the Eq. 10.4 to the ensemble of particles with various sizes. Particles with radiuses in range r, $r + dr$ do yield to the volume extinction coefficient $(N(r+dr) - N(r))C_e(r) = n(r)C_e(r)dr$. The summation of these contributions including Eqs. 10.1 and 10.2 gives the integral

$$\alpha = N\int_0^\infty f(r)C_e(r,m)dr \tag{10.5}$$

where $C_e(r,m)$ is the cross-section of the particle with the radius r and particle matter CRI m. It is to point out that there is in Eq. 10.4 and in other similar expressions the dependence of wavelength λ via the parameter $y = 2\pi r/\lambda$ including in Mie formulas.

The analogues relation is derived for the volume scattering coefficient σ:

$$\sigma = N\int_0^\infty f(r)C_s(r,m)dr \tag{10.6}$$

where $C_s(r,m)$ is the cross-section of the single particle. And the similar for the volume absorption coefficient κ

$$\kappa = \alpha - \sigma = N\int_0^\infty f(r)C_e(r,m)dr - N\int_0^\infty f(r)C_s(r,m)dr \tag{10.7}$$

The summation principle for the phase function is formulated as the product of the directed scattering cross-section and the function distribution. The expression for calculating the phase function of the aerosol ensemble is the following

$$x(\gamma) = \frac{1}{\sigma} \int_{0}^{\infty} f(r) C_s(r,m) x(\gamma, r, m) dr \quad (10.8)$$

where the ensemble phase function is on the left and the phase function of single particle is under the integral.

Thus, relations (10.5)–(10.8) provide calculating optical characteristics of atmospheric aerosols with function of size distribution *f(r)*.

10.5 Practice 9

10.5.1 Objectives

The purpose of the practice is the study of the dependence of the atmospheric aerosols optical characteristics on parameters of the set of distribution functions.

Let the aerosol concentration $N = 1\ \text{cm}^{-3}$ because of the trivial dependence (direct proportionality) of the considered characteristics on concentration N, that is all the calculations shall be carried out for a unit concentration of particles.

Two dependencies of characteristics are considered as the most interesting for educational points:

1. The dependence of volume coefficients of aerosol extinction (10.5), scattering (10.6) and absorption (10.7) on wavelength.
2. The aerosol phase functions against scattering angle γ (10.8) at fixed wavelength (0.55 μm).

The following typical atmospheric aerosols are taken for consideration: soot (the major component of anthropogenic aerosols), dust (the major component of continental aerosols), water (clouds, fogs, mist, precipitations). Tables of complex refractive indexes are in the attached file "*cri3.dt*".

10.5.2 Sequential Steps of the Exercise Implementation

The computer program "AEROPTIC.exe" realizes the above algorithm for calculation. After execution the program runs in the dialogue regime. It is necessary for the file "*cri3.dt*" to be in the same directory for the program to operate.

After execution and giving out information, the first question of the program is:
Input filename for results:

A name of the resulting file is input (entered). The initial parameters and output results are in the file at every moment the program is running. The file is open for writing till program operation ends.

Then the program fulfills two stages of running.

1. The first is studying the dependences of aerosol volume coefficients of extinction, scattering, and absorption on wavelength.
   Sequentially the following questions are output at the screen:
   Input size distribution function parameters (r0 mkm and s) for soot aerosol:
   It is recommended to input the typical values $r_0 = 0.05\_s = 1.1$ (values are separated by space between them), then press "Enter".
   It is possible to input several couples of values. They divided with spaces and at the end press "Enter". For example, to input two variants 0.05, 1.1 *and* 0.1, 2.5, enter the four values separate by spaces 0.05_1.1_0.1_2.5 and then "Enter".
   Input size distribution function parameters (r0 mkm and s) for dust aerosol:
   The two-tree couples of typical values $r_0 = 0.1\_s = 0.6$ (Aitken nuclei), $r_0 = 0.5\_s = 0.8$ (Large aerosol), $r_0 = 2.0$, $s = 1.0$ (Giant aerosol). Rules for inputting values are similar to the precedent item.
   Input size distribution function parameters (r0 mkm and s) for water aerosol:
   Typical values are $r_0 = 0.5$, $s = 0.4$ (drizzles in clouds); $r_0 = 3.0$, $s = 0.6$ (small droplets in stratus cloud), $r_0 = 10.0$, $s = 0.8$ (large droplets in cumulus cloud).
   The program calculates volume coefficients of extinction, scattering, and absorption as a function of the wavelength. Results are output in the resulting file.
2. The second stage is the studying of aerosol phase function as a function scattering angle.
   The following questions are output at the screen sequentially:

   Input size distribution function parameters (r0 mkm and s) for soot aerosol:
   Input size distribution function parameters (r0 mkm and s) for dust aerosol:
   Input size distribution function parameters (r0 mkm and s) for water aerosol:

Responses are similar to those discussed above. Giant water droplets and rain droplets are recommended to add with parameters $r_0 = 200.0\_s = 0.5$.

The program calculates the phase function as a function of the scattering angle at the wavelength 0.55 μm and output results in the file.

Then the following is needed:

- Plotting the dependencies of volume coefficients of extinction, scattering and absorption against the wavelength. It is better to take the logarithmic scale of the ordinate axe. The dependences at shortwave and longwave ranges are to be plotted separately.
- Describing (and physically interpreting if possible) obtained dependences.
- From the results at the second stage the following is needed:
- Plotting phase function as a function of the scattering angle. The logarithmic scale of ordinate axe is needed.
- Describing and physically interpreting the variations of phase function shapes while increasing mean radius.

Elucidate the following questions in the concise report: Is there the Rayleigh phase function (for which particles and why)? Are there local maximums at angles larger than 90° (that are responsible for rainbow)?

Chapter 11
Calculating Solar Radiative Characteristics in Clouds with Asymptotic Formulas of the Radiative Transfer Theory

Abstract Asymptotic formulas of the radiative transfer theory are presented for thick atmosphere with weak absorption. The applicability region over the optical thickness, single scattering albedo, and phase function asymmetry parameter is discussed together with the approach exactness. The simplest model of cloud layer is proposed for calculation of radiative characteristics.

11.1 The Basic Formulas

Let us consider the model of an extended and horizontally homogeneous cloud of a big optical thickness $\tau \gg 1$ as Fig. 11.1 illustrates. Here, the cloud layer is assumed vertically homogeneous as well and the influence of the clear atmosphere layers above and below the cloud layer is not taken into account. The volume coefficients of scattering α and absorption κ, linked with the cloud characteristics as $\kappa + \alpha \equiv \tau_0/\Delta z$, $\alpha \equiv \omega_0\tau_0/\Delta z$, $\kappa \equiv \tau_0(1 - \omega_0)/\Delta z$, are used for the cloud description. The optical properties of the cloud are described by the following parameters: single scattering albedo (or the probability of photon surviving in a single interaction act) ω_0; optical thickness τ, and mean cosine of the scattering angle (asymmetry factor) g, which characterizes the Henyey-Greenstein phase function (Eq. 1.16, Chap. 1). From the bottom the cloud layer adjoins the ground surface and its reflectance is described by ground albedo A. The underlying atmosphere could be taken into account if albedo A is implying as an albedo of the system "surface + atmosphere under the cloud". Parallel solar flux F_0 is falling on the cloud top at incident angle arccosμ_0. The reflected and transmitted radiance is observed at viewing angle arccosμ. The reflected radiance (in the units of incident extraterrestrial flux $F_0\mu_0$) is expressed with reflection function $\rho(\tau,\mu,\mu_0)$ and the transmitted radiance (in the same units) is expressed with transmission function $\sigma(\tau,\mu,\mu_0)$.

At the sufficiently big optical depth within cloud layer far enough from the top and bottom boundaries the asymptotic or diffusion regime is installed owing to the multiple scattering. This regime permits a rather simple mathematical description.

I. Melnikova et al., *Remote Sensing of the Environment and Radiation Transfer*,
DOI 10.1007/978-3-642-14899-6_11,

Fig. 11.1 The model of the atmosphere

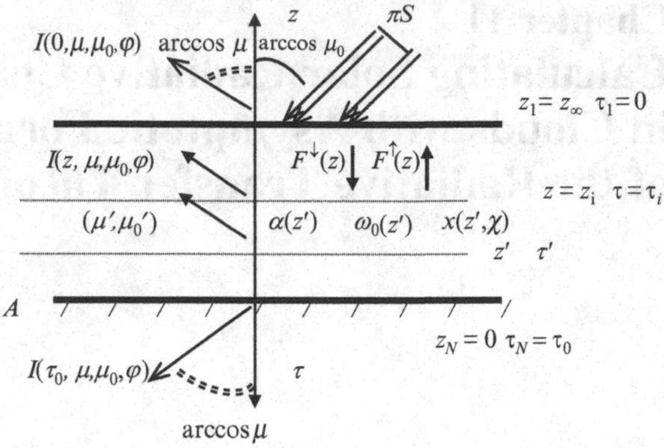

The region within cloud layer is called a diffusion domain. The physical meaning yields the following specific features of the diffusion domain:

1. the role of the direct radiation (transferred without scattering) is negligibly small comparing to the role of the diffused radiation;
2. the radiance within the diffusion domain does not depend on the azimuth;
3. the relative angle distribution of the radiance does not depend on the optical depth.

The name "diffusion" is appearing because the equation of radiative transfer is transformed to the diffusion equation in that case.

Remember the equation system (1.29) derived in the Chap. 1 and assume the approximation:

$$\int_{-1}^{1} I(\tau,\mu)\mu^2 d\mu = D \int_{-1}^{1} I(\tau,\mu) d\mu,$$

This relation is the strict in an inner domain remote from boarders of the optically thick cloud $K(\tau,\mu) = D\, I(\tau,\mu)$. The value D is called the diffuse constant. It was shown that in the scattering layer of a big optical thickness the analytical solution is expressed through the asymptotic formulas of the radiative transfer theory, moreover the existence and uniqueness of the solution have been proved. It is expressed through reflection $\rho(\tau,\mu,\mu_0)$ and transmission $\sigma(\tau,\mu,\mu_0)$ functions of multiple scattered radiation, in the following:

$$\begin{aligned}\rho(\tau,\mu,\mu_0,\varphi) &= \rho_\infty(\mu,\mu_0,\varphi) - \frac{ml\bar{K}(\mu)K(\mu_0)\exp(-2k\tau)}{1 - l\bar{l}\exp(-2k\tau)}\\ \sigma(\tau,\mu,\mu_0) &= \frac{m\bar{K}(\mu)K(\mu_0)\exp(-k\tau)}{1 - l\bar{l}\exp(-2k\tau)},\end{aligned} \quad (11.1)$$

In these equations $\rho_\infty(\mu,\mu_0,\varphi)$ is the reflection function for a semi-infinite atmosphere; $K(\mu)$ is the escape function, which describes an angular dependence of the reflected and transmitted radiance; m, l, k are the constants, depending on the cloud optical properties, the formulas for its computing are presented below; $\bar{K}(\mu)$ and $\bar{l}$ depends on surface albedo A as well. The following expressions are taking into account the ground reflection:

$$\begin{gathered}\bar{K}(\mu) = K(\mu) + A\bar{n}a(\mu)\\ \bar{n} = n/(1 - Aa^\infty)\\ \bar{l} = l - Am\bar{n}n,\end{gathered} \tag{11.2}$$

where the plane albedo $a(\mu)$ and the spherical albedo a^∞ of the infinite atmosphere, and the value n are defined by integrals:

$$\begin{gathered}a(\mu) = 2\int_0^1 \rho(\mu,\mu_0)\mu_0 d\mu_0\\ a^\infty = 2\int_0^1 a(\mu)\mu d\mu\\ n = 2\int_0^1 K(\mu)\mu d\mu, \quad \bar{n} = 2\int_0^1 \bar{K}(\mu)\mu d\mu,\end{gathered} \tag{11.3}$$

It is seen that Eq. 11.1 are asymmetric relatively to variables μ and μ_0, which are input with escape functions $K(\mu)$ and $\bar{K}(\mu)$. It links with different boundary conditions at the top and bottom of the layer. The top is free and it could be assumed as an absolutely absorbing one for the upward radiation and the bottom boundary reflects partly (1-A) the downward radiation. Thus each of them generates its own light regime described by different asymptotic functions $K(\mu)$ and $\bar{K}(\mu)$ and constants l and $\bar{l}$.

Consider the semispherical fluxes of diffused solar radiation (solar irradiances) in relative units of incident solar flux F_0. Reflected irradiance $F^\uparrow(0,\mu_0)$ and transmitted irradiance $F^\downarrow(\tau,\mu_0)$ are described by the formulas similar to Eq. 11.1, where reflection function $\rho_\infty(\mu,\mu_0)$ and escape function $K(\mu)$ are substituted with their integrals $a(\mu_0)$ and n, according to Eqs. 11.2 and 11.3. As a result, the following formulas are inferred:

$$\begin{aligned}F^\uparrow(0,\mu_0) &= a(\mu_0) - \frac{mn\bar{l}K(\mu_0)\exp(-2k\tau)}{1 - l\bar{l}\exp(-2k\tau)}\\ F^\downarrow(\tau,\mu_0) &= \frac{m\bar{n}K(\mu_0)\exp(-k\tau)}{1 - l\bar{l}\exp(-2k\tau)}.\end{aligned} \tag{11.4}$$

The radiation absorption within the cloud layer is determined by the radiative divergence Eq. (1.5). It is computed with the obvious equation:

$$R = 1 - F^{\uparrow}(0,\mu_0) - (1-A)F^{\downarrow}(\tau_0,\mu_0)$$
$$= 1 - a(\mu_0) + \frac{nK(\mu_0)me^{-k\tau_0}}{1 - l\bar{l}e^{-2k\tau_0}}\left[\bar{l}e^{-k\tau_0} - \frac{1-A}{1-Aa_\infty}\right]. \quad (11.5)$$

Mention that the term "asymptotic" specifies the light regime installed within the cloud and it does not point out any approximation. Equations 11.1, 11.4 and 11.5 are strict within the diffusion domain. Their accuracy will be studied below depending on the optical thickness.

11.2 Weak True Absorption of Solar Radiation

In clouds, the absorption is extremely weak compared to scattering (1 $\omega_0 < <1$) within the short-wavelength range. As it has been shown in books in this case both functions $\rho_\infty(\mu,\mu_0)$ and $K(\mu)$ and constants m, l, k are expressed with the expansions over powers of small parameter $(1 - \omega_0)$. We consider here that parameter s, where $s^2 = (1 - \omega_0)/[3(1 - g)]$, is more convenient for the problem in question than parameter $(1 - \omega_0)$. Value g is a mean cosine of the scattering angle or, here, the parameter of Henyey-Greenstein function (1.15). Then, these expansions over the powers of s (in case of keeping the terms with the power equal to two) for the constants in Eqs. 11.1–11.5 are looking as follows:

$$k = 3(1-g)s\left[1 + s^2\left(1.5g - \frac{1.2}{1+g}\right)\right]$$
$$m = 8s\left[1 + \left(6 - 7.5g + \frac{3.6}{1+g}\right)s^2\right]$$
$$l = 1 - 6q's + 18q'^2s^2 \quad (11.6)$$
$$a^\infty = 1 - 4s + 12q's^2 - \left(36q' - 6g - \frac{1.608}{1+g}\right)s^3$$
$$n = 1 - 3q's + \left(9q'^2 - 3(1-g) - \frac{2}{1+g}\right)s^2.$$

For the functions in Eqs. 11.1–11.5 the followings expansions are correct:

$$K(\mu) = K_0(\mu)(1 - 3q's) + K_2(\mu)s^2$$
$$a(\mu) = 1 - 4K_0(\mu)s + a_2(\mu)s^2 + a_3(\mu)s^3 \quad (11.7)$$
$$\rho_\infty(\mu,\mu_0) = \rho_0(\mu,\mu_0) - 4K_0(\mu)K_0(\mu_0)s + \rho_2(\mu,\mu_0)s^2 + \rho_3(\mu,\mu_0)s^3,$$

where the nomination is introduced: $q' = 2\int_0^1 K_0(\zeta)\zeta^2 d\zeta \cong 0.714$

In these expansions functions $\rho_0(\mu,\mu_0)$ and $K_0(\mu)$ are functions $\rho_\infty(\mu,\mu_0)$ and $K(\mu)$ for the conservative scattering ($\omega_0 = 1$) correspondingly, functions $a_2(\mu)$ and $K_2(\mu)$ are the coefficients by the item s^2. They are presented either in analytical or in table form. Asymptotic expansions (11.6) and (11.7) have been mathematically rigorously derived, their errors are defined by items ~ s^3 or ~ s^4 omitted in the series.

The coefficients of items s^2 and s^3 in the expansion for reflection function $\rho_\infty(\mu,\mu_0)$ are looking as:

$$\rho_2(\mu,\mu_0) = \frac{a_2(\mu)a_2(\mu_0)}{a_2}, \quad \rho_3(\mu,\mu_0) = \frac{a_3(\mu)a_3(\mu_0)}{a_3}, \tag{11.8}$$

where a_2, a_3, $a_2(\mu)$ and $a_3(\mu)$ are the coefficients of s^2 and s^3 in the series for spherical a^∞ albedo as per Eq. 11.6 and in series for plane $a(\mu)$ albedo as per Eq. 11.7 respectively. Values of the conservative escape function $K_0(\mu_0)$ are presented in the following table (Table 11.1):

The approximation for function $K_0(\mu)$ with the error 3% for $\mu > 0.4$ has been proposed as: $K_0(\mu) = 0.5 + 0.75\,\mu$. The analysis of numerical results yields the following approximation for coefficients $K_0(\mu)$ and $K_2(\mu)$ with taking into account the phase function dependence:

$$K_0(\mu) = (0.7678 + 0.0875g)\mu + 0.5020 - 0.0840g. \tag{11.9}$$

$$K_2(\mu) = n_2 K_0(\mu) w(\mu) = 1.667 n_2(\mu^2 + 0.1)$$

The correlation coefficient of the formulas (11.9) and numerical calculations is about 0.99 – 0.93 depending on parameter g.

Expressions for functions $a_2(\mu)$, and $a_3(\mu)$ are follows:

$$\begin{aligned} a_2(\mu) &= 3K_0(\mu)\left(\frac{3}{1+g}(1.271\mu - 0.9) + 4q'\right) \\ a_3(\mu) &= 4K_0(\mu)\left[4.5g - \frac{1.6}{1+g} - 3 - n_2 w(\mu)\right]. \end{aligned} \tag{11.10}$$

The values of function $a_2(\mu)$ are presented in Table 11.2 for four values of parameter g.

Table 11.1 The escape function $K_0(\mu_0)$ for clouds ($0.65 \le g \le 0.9$)

μ_0	1.0	0.9	0.8	0.7	0.6	0.5	0.4	0.3	0.2	0.1
$K_0(\mu_0)$	1.271	1.193	1.114	1.034	0.952	0.869	0.782	0.690	0.591	0.476

Table 11.2 Values of the second coefficient $a_2(\mu)$ of the plane albedo expansion

g \ μ	0	0.1	0.2	0.3	0.4	0.5	0.6	0.7	0.8	0.9	1.0
0.75	1.310	2.220	3.118	4.078	5.126	6.256	7.475	8.786	10.19	11.70	13.29
0.80	1.267	2.236	3.151	4.117	5.163	6.289	7.494	8.796	10.18	11.66	13.23
0.85	1.201	2.242	3.181	4.148	5.198	6.320	7.512	8.798	10.17	11.63	13.18
0.90	1.092	2.244	3.208	4.193	5.237	6.350	7.529	8.808	10.16	11.60	13.12

Surface albedo A is assumed by formulas:

$$\bar{K}_0(\mu) = K_0(\mu) + A/(1-A),$$
$$\bar{K}_2(\mu) = K_2(\mu) + \frac{A}{1-A}\left[3K_0(\mu)\frac{3.8\mu - 2.7}{1+g} + n_2\right], \qquad (11.11)$$

11.3 The Analytical Presentation of the Reflection Function

The following group of the formulas is the approximations obtained from the analysis of the numerical values of the reflection function. As it is usually done, let us describe the reflection function over the azimuth angle cosine to separate the item independent of the azimuth angle:

$$\rho(\varphi, \mu, \mu_0) = \rho^0(\mu, \mu_0) + 2\sum_{m=1}^{\infty} \rho^m(\mu, \mu_0)\cos m\varphi, \qquad (11.12)$$

where functions $\rho^m(\mu,\mu_0)$ are the harmonics of the reflection function of order m. Superscripts specify here the number of the azimuthal harmonics. As it has been mentioned above, we are using here the phase function described by Henyey-Greenstein formula (1.15).

The analysis of numerical calculations shows that for the accurate description of function $\rho(\varphi,\mu,\mu_0)$ it is enough to know the zeroth and first six harmonics if either of cosines μ and μ_0 are greater than 0.15 even for value $g = 0.9$ unfavorable for the computing accuracy. This limitation does not restrict our consideration because it is also necessary to use a complicated model of the spherical atmosphere and to take into account the refraction of solar rays for the small cosines of zenith solar and viewing angles. These cases are not studied here.

The values of $\rho^m(\mu,\mu_0)$ for m = 0,. . .,6 have been analyzed and the following expression, is used for the description of high harmonics $\rho^m(\mu,\mu_0)$:

$$\rho^m(\mu, \mu_0) = [a^m\mu\mu_0 + b^m(\mu + \mu_0) + c^m]/(\mu + \mu_0) \qquad (11.13)$$

This presentation provides the reciprocity of the reflection function relatively to the both zenith viewing and zenith solar angles.

Table 11.3 Linear approximation for coefficients a^m^, b^m^ и c^m^ in formula (7.11) for zero, first and second azimuthal harmonics of the reflection function

m	a^m	b^m	c^m	μ_{limit}
0	$2.051\ g + 0.508$	$-1.420\ g + 0.831$	$0.930\ g + 0.023$	–
1	$1.821\ g - 0.558$	$-1.413\ g + 0.387$	$1.150\ g - 0.239$	0.80
2	$2.227\ g - 0.669$	$-1.564\ g + 0.481$	$1.042\ g - 0.293$	0.55

Table 11.4 Power approximation for the coefficients a^m^, b^m^ и c^m^ in Eq. (2.14) for 3rd, 4th, 5th and 6th azimuthal harmonics of the reflection function

m	$0.3 \le g \le 0.9$			
	a^m	b^m	c^m	η_{limit}
3	$62.00\ g^3 - 90.28\ g^2 + 42.42\ g - 6.26$	$-15.24\ g^3 + 19.70\ g^2 - 8.73\ g + 1.25$	$2.75\ g^2 - 2.03\ g + 0.39$	0.50
4	$105.26\ g^3 - 155.06\ g^2 + 72.93\ g - 10.76$	$-30.30\ g^3 + 43.04\ g^2 - 19.83\ g + 2.89$	$3.70\ g^2 - 3.20\ g + 0.65$	0.45
5	$120.63\ g^3 - 177.60\ g^2 + 83.48\ g - 12.32$	$-25.84\ g^3 + 35.15\ g^2 - 15.61\ g + 2.22$	$3.23\ g^2 - 2.75\ g + 0.55$	0.35
6	$144.92\ g^3 - 202.16\ g^2 + 90.48\ g - 12.85$	$-32.60\ g^3 + 43.88\ g^2 - 19.15\ g + 2.67$	$3.90\ g^2 - 3.41\ g + 0.70$	0.35

The approximation of coefficients a^m, b^m and c^m in the range of parameter g $0.3 \le g \le 0.9$ is presented in Tables 11.3 and 11.4.

The well-known relation of the rigorous theory is assumed for the isotropic and conservative scattering ($g = 0$, $\omega_0 = 1$), namely:

$$\rho^0(\mu, \mu_0) = \frac{\varphi(\mu)\varphi(\mu_0)}{4(\mu + \mu_0)}, \tag{11.14}$$

where $\varphi(\mu)$ is Ambartsumyan's function. In this case the following approximation is correct: $\varphi(\mu) = 1.874\ \mu + 1.058$ and it has been obtained that $a^0 = 0.88$, $b^0 = 0.47$, and $c^0 = 0.28$. It is known that the reflection function for the isotropic scattering does not differ very much from the anisotropic values of $\rho^0(\mu,\mu_0)$ if μ, $\mu_0 > 0.25$, so it is possible to improve this approach for the enlarged angle ranges. The simple formula for the isotropic scattering could be corrected approximately with the linear dependence of the phase function parameter as follows:

$$\rho^0(\mu, \mu_0) = \frac{\varphi(\mu)\varphi(\mu_0) + g[4.8\mu_0\mu - 3.0(\mu_0 + \mu) + 1.9]}{4(\mu_0 + \mu)}. \tag{11.15}$$

In the case of Henyey-Greenstein phase function the high harmonics are close to zero ($\rho^m(\mu,\mu_0) \approx 0$, m > 0) if either of zenith angle cosines μ and μ_0 are greater than μ_{limit}. The values of μ_{limit} are different for different harmonics and they are shown in Tables 2.3 and 2.4. The approximation by Eq. 11.13 with coefficients a^m,

b^m and c^m in Tables 11.3 and 11.4 gives an acceptable presentation for all the harmonics of the reflection function considered here. The errors of this approximation have been shown to depend on the values of the zenith solar and viewing angles cosines, on number of the harmonic *m*, and on phase function parameter *g*. Some details of the error analysis will be presented in Sect. 11.4.

The presented asymptotic formulas (11.1), (11.4) and (11.5), expansions (11.6), (11.7) and approximations (11.8)–(11.11), (11.13) and (11.15) allow computing the reflected and transmitted radiance and irradiance together with the radiative divergence for cloud layer if the layer properties and the geometry of the problem are known. The considered model has to satisfy the applicability ranges of the presented formulas: *big optical thickness* and *weak true absorption*. These ranges will be analyzed in Sect. 11.4 in detail. However, it is necessary to point out that for the application of Eqs. 11.1–11.5 the big optical thickness with known asymptotic functions and constants is enough. The use of the expansions (11.6) and (11.7) needs the weak absorption.

11.4 Diffused Radiation Field Within the Cloud Layer

Radiation within the cloud layer (in the diffusion domain: $\tau_0 - \tau_{N\text{-}1} \gg 1$ and $\tau_1 \gg 1$) is described with formulas different from those presented above. Here we are offering the results useful for the further consideration.

The diffused radiance in energetic units in the diffusion domain at optical depth τ satisfied conditions $\tau >> 1$, $\tau_0 - \tau >> 1$ is expressed with the equation:

$$I(\tau,\mu,\mu_0,\tau_0) = \frac{F_0}{\pi} K(\mu_0)\mu_0 \mathrm{e}^{-k\tau_0} \frac{i(\mu)\mathrm{e}^{k(\tau_0-\tau)} - i(-\mu)\bar{l}\mathrm{e}^{-k(\tau_0-\tau)}}{1 - l\bar{l}\mathrm{e}^{-2k\tau_0}}, \tag{11.16}$$

where F_0 is the solar incident flux, function $i(\mu)$ characterizes the angle dependence of the radiance in deep levels of the semi-infinite atmosphere. The behavior of function $i(\mu)$ relatively the phase function shape and absorption in the medium has been studied. The expansion for function $i(\mu)$ has been derived in case of the weak true absorption, which is presented here in terms of parameter *s*:

$$\begin{aligned} i(\mu) &= 1 + 3s\mu + 3\frac{1-g^2+2\mathrm{P}_2(\mu)}{1+\mathrm{g}}s^2 \\ &+ \left[9(1-1.5g)\mu + \frac{10.8P_3(\mu)}{(1+g)(1+g+g^2)} + \frac{3.6\mu}{1+g}\right]s^3 \end{aligned} \tag{11.17}$$

Functions $P_i(\mu)$ for $i = 1,2, \ldots$ are Legendre polynomials of power *i*.

The diffused irradiance in relative units of F_0 within the optically thick cloud layer is described with the following:

$$F^{\downarrow}(\mu_0,\tau,\tau_0)=\frac{K(\mu_0)\mathrm{e}^{-\mathrm{k}\tau_0}}{1-l\bar{l}\mathrm{e}^{-2\mathrm{k}\tau_0}}\left[i^{\downarrow}\mathrm{e}^{\mathrm{k}(\tau_0-\tau)}-i^{\uparrow}\bar{l}\mathrm{e}^{-\mathrm{k}(\tau_0-\tau)}\right],$$
$$F^{\uparrow}(\mu_0,\tau,\tau_0)=\frac{K(\mu_0)\mathrm{e}^{-\mathrm{k}\tau_0}}{1-l\bar{l}\mathrm{e}^{-2\mathrm{k}\tau_0}}\left[i^{\uparrow}\mathrm{e}^{\mathrm{k}(\tau_0-\tau)}-i^{\downarrow}\bar{l}\mathrm{e}^{-\mathrm{k}(\tau_0-\tau)}\right], \tag{11.18}$$

where $i^{\downarrow}=2\int\limits_0^1 i(\mu)\mu d\mu,\quad i^{\uparrow}=2\int\limits_0^1 i(-\mu)\mu d\mu.$

Expansions for values $i^{\downarrow}$ and $i^{\uparrow}$ are derived after integrating Eq. 11.17:

$$i^{\downarrow\uparrow}=1\pm 2s+3s^2\frac{1.5-g^2}{1+g}\pm 3s^3\left[2-3g+\frac{0.8}{1+g}\right] \tag{11.19}$$

It is also convenient to describe the internal radiation field with the values of *internal albedo* $b(\tau_i)=F^{\uparrow}(\tau_i)/F^{\downarrow}(\tau_i)$ and net flux $F(\tau_i)=F^{\downarrow}(\tau_i)-F^{\uparrow}(\tau_i)$:

$$F(\tau,\mu_0)=F^{\downarrow}(\tau,\mu_0)-F^{\uparrow}(\tau,\mu_0)=\frac{4sK(\mu_0)\mathrm{e}^{-k\tau}}{1-l\bar{l}\mathrm{e}^{-2k\tau_0}}\left[1+\bar{l}\mathrm{e}^{-2k(\tau_0-\tau)}\right]$$
$$\frac{F^{\uparrow}(\tau,\mu_0)}{F^{\downarrow}(\tau,\mu_0)}=b(\tau)=\frac{b^{\infty}-\bar{l}\mathrm{e}^{-2k(\tau_0-\tau)}}{1-b^{\infty}\bar{l}\mathrm{e}^{-2k(\tau_0-\tau)}} \tag{11.20}$$

Value b^{∞} and function $b(\tau)$ are called the internal albedo of the infinite atmosphere and the internal albedo of the atmosphere of the big optical thickness respectively, moreover $b^{\infty}=1–4s+8s^2$ and the values of function $b(\tau)$ could be obtained from the observations or from the calculations of the semispherical irradiances at level τ.

11.5 Case of the Conservative Scattering

In the true absorption absence, according the definition, we have $\omega_0=1$ and the expressions for the radiation characteristics are particularly simple.

For the reflection and transmission functions:

$$\rho(0,\mu,\mu_0,\varphi)=\rho_0(\mu,\mu_0,\varphi)-\frac{4K_0(\mu_0)K_0(\mu)}{3\left[(1-g)\tau_0+2q'+\frac{4A}{3(1-A)}\right]},$$
$$\sigma(\tau_0,\mu,\mu_0)=\frac{4K_0(\mu_0)\bar{K}_0(\mu)}{3\left[(1-g)\tau_0+2q'+\frac{4A}{3(1-A)}\right]}; \tag{11.21}$$

for the semispherical fluxes in relative units of F_0

$$F^{\uparrow}(0,\mu_0) = 1 - \frac{4K_0(\mu_0)}{3\left[(1-g)\tau_0 + 2q' + \frac{4A}{3(1-A)}\right]},$$
$$F^{\downarrow}(\tau_0,\mu_0) = \frac{4AK_0(\mu_0)}{3(1-A)\left[(1-g)\tau_0 + 2q' + \frac{4A}{3(1-A)}\right]}, \quad (11.22)$$

and, finally, the simple expression for the net flux that summarizes both Eq. 11.22 is feasible at any level in the conservative medium because the net flux is constant without absorption

$$F(\tau,\mu_0) = \frac{4K_0(\mu_0)(1-A)}{3(1-A)[(1-g)\tau_0 + \delta] + 4A}. \quad (11.23)$$

It should be emphasized that the equality $F^{\uparrow}(\tau,\mu_0) = F^{\downarrow}(\tau,\mu_0) = K_0(\mu_0)$ is correct in the semi-infinite conservatively scattered atmosphere at the big optical depth, where the sense of the escape function $K_0(\mu)$, frequently met in our consideration, is clear from. The case of the conservative scattering comes true in a certain cloud layer at the some wavelengths within the visual spectral range. Equations 11.21–11.22 are correct in the wider interval of the optical depth ($\tau_0 \geq 3$) than Eqs. 11.1 and 11.4 derived with taking into account the absorption. Corresponding relations of the characteristics of the inner radiation field are written as:

for the radiance:

$$I(\tau,\mu) = \frac{F_0\mu_0K_0(\mu_0)\{(1-A)[3(1-g)(\tau_0-\tau) + 3q' + 3\mu] + 4A\}}{\pi\{(1-A)[3(1-g)\tau_0 + 6q'] + 4A\}}, \quad (11.24)$$

for the upward and downward semispherical solar fluxes:

$$F^{\uparrow}(\tau,\mu_0) = K_0(\mu_0)\frac{(1-A)[3(1-g)(\tau_0-\tau) + 3q' - 2] + 4A}{(1-A)[3(1-g)\tau_0 + 6q'] + 4A}$$
$$F^{\downarrow}(\tau,\mu_0) = K_0(\mu_0)\frac{(1-A)[3(1-g)(\tau_0-\tau) + 3q' + 2] + 4A}{(1-A)[3(1-g)\tau_0 + 6q'] + 4A}. \quad (11.25)$$

The formulas of the radiation characteristics in case of conservative scattering are possible to apply for the rough estimation even for the very weak absorption but the computational errors increase fast when the absorption grows and it is necessary to use the equations for the absorption medium to reach certain accuracy.

11.6 Error Analysis

Asymptotic formulas of the radiation transfer theory presented in this chapter are obtained strictly. It is necessary to take into consideration that they are describing the radiation field within the cloud layer and at the cloud top and base boundaries the more exact, the bigger the optical thickness and the weaker true absorption.

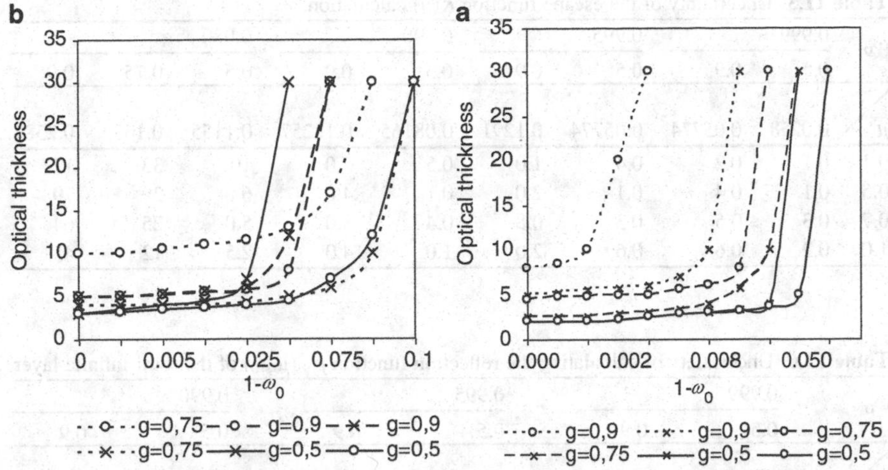

Fig. 11.2 The applicability ranges of the asymptotic formulas of the radiative transfer theory in case of calculation reflected irradiance (**a**) and radiative divergence (**b**) for cloudy layer. *Curves corresponded to the relative uncertainty equal to 5%. Solid curve is a phase function parameter* $g = 0.5$*; dashed line –* $g = 0.75$ *and dashed dotted line –* $g = 0.9$*; curves with circles correspond to* $\mu = 1$*, with crosses – to* $\mu = 0.5$

In addition, there is a strong dependence of the accuracy upon the degree of the scattering anisotropy (the extension forward of the phase function defined by the parameter g magnitude). The uncertainties of formulas for reflected and transmitted radiation are about 2% beginning from optical thickness $\tau_0 \geq 4/(1-k)$. The numerical analysis of formulas for the spherical albedo and radiation transmittance for the wide set of parameters shows uncertainty not exceed 5% by values $\tau \geq 2.0$ and $\omega_0 \geq 0.7$.

The accuracy of formulas for irradiances was tested with closed numerical experiments to provide relative errors less than 5% in the region plotted in coordinates "$\tau - \omega_0$" in Fig. 11.2. Curves in Fig. 11.2a, b correspond to the level of 5% error of the reflected irradiance (a) and radiative divergence (b) calculated for parameters $g = 0.5$; 0.75 and 0.9 and for two values of cosine $\mu_0 = 1$ and 0.5.

The numerical analysis of the accuracy of the radiance calculation within the optically thick layer gives the applicability region of the radiance ($\tau > 15$; $\omega_0 > 0.99$) is restricted stronger than of the irradiance ($\tau > 7$; $\omega_0 > 0.9$), which in turn is narrower than for spherical albedo and transmittance ($\tau > 2$; $\omega_0 > 0.8$).

Errors of asymptotic expansions (11.6) and (11.7) are defined by omitted items proportional to s^3 or s^4. The accuracy of the approximations was tested by comparison with the function values computed by the numerical methods. The relative uncertainties of the escape function computed with approximations (11.8) are presented in Table 11.5. It has been found that uncertainties are rather small as far as $\omega_0 = 0.98$ for magnitudes $g = 0.5$ and $\mu > 0.2$. Table 11.5 are illustrating that the errors of the escape function calculation are not exceeding 6% for value $s < 0.12$.

Table 11.5 Uncertainty of the escape function $K(\mu)$ calculation,%

ω_0	0.999		0.995		0.990		0.980		
g	0.5	0.9	0.5	0.9	0.5	0.9	0.5	0.75	0.9
μ \ s	0.0258	0.05774	0.05774	0.1291	0.08165	0.18257	0.1155	0.1633	0.2582
0.1	0.1	0.2	0.4	1.0	0.5	2.0	10	33	127
0.5	0.1	0.4	0.1	2.0	0.1	4.0	6.0	29	79
0.7	0.3	0.5	0.3	0.8	0.4	3.0	5.0	25	64
1.0	0.2	0.6	0.6	2.0	1.0	4.0	2.5	12	45

Table 11.6 Uncertainty of calculating the reflection function $\rho_\infty(\mu,\mu_0)$ of the semi-infinite layer

ω_0	0.999		0.995		0.990	
g	0.5	0.9	0.5	0.9	0.5	0.9
μ \ s	0.0258	0.05774	0.05774	0.1291	0.08165	0.18257
0.1	0.2	0.6	0.2	1.0	0.3	2.6
0.5	0.2	0.3	0.4	1.0	1.0	3.0
1.0	0.2	0.3	0.5	1.0	0.7	3.0

Comparison of reflection function $\rho_\infty(\mu,\mu_0)$ calculation results accounting coefficients $\rho_2(\mu,\mu_0)$ and $\rho_3(\mu,\mu_0)$ with numerical computing results yields the errors shown in Table 11.6. Equation 11.8 for functions $\rho_2(\mu,\mu_0)$ and $\rho_3(\mu,\mu_0)$ allow computing the corresponded values with rather small error as far as $\omega_0 = 0.9$. Therefore, the solar radiance reflected from the cloud layer in the shortwave spectral range is possible to calculate with the analytical formulas, and this fact is useful for the interpretation of the satellite radiation data.

11.6.1 Optical Model of the Cloud Layer

The input of optical parameters is necessary for calculating radiative characteristics Ranges of cloud optical parameters are presented below:

1. The optical thickness $100 \geq \tau_0 \geq 5$;
2. The optical depth within the layer τ is automatically assumed with dividing the layer to five parts;
3. The single scattering albedo $1 \geq \omega_0 \geq 0.992$;
4. The asymmetry parameter of the phase function $g = 0.5$–0.85;
5. The surface albedo $A = 0$–0.95
6. Solar and viewing zenith angles (degrees) 0–80
7. The azimuth angle relative to the Sun φ (degrees) for radiance calculating 0–180

11.7 Practice 10

11.7.1 Objectives

Calculating radiative characteristics (diffuse irradiance outgoing from cloud layer, diffuse irradiance within layer at four levels, diffuse radiance at boarders and within cloud layer and radiative divergence) for the input optical model and geometry of illuminating. It is to study variations of irradiances and radiative divergence versus the optical model (optical thickness, single scattering albedo, phase function parameter, surface albedo) and solar zenith angle, to plot obtained dependencies and to prepare the report.

11.7.2 Sequential Steps of the Exercise Implementation

Calculation is accomplished with compute program compiled in Borland 4 C++. The file "ASYMP.exe" propose the following dialog when questions are output at the screen sequentially.

1. Input number of wavelength
   It is recommended to input the number of calculation variants;
   If the dependency on one parameter is studied (e.g. optical thickness) the set of several (e.g. 10) parameter values is input and other parameters are repeated.
2. Input number of viewing zenith angles
   It is recommended to input the number of viewing angles if the intensity is calculated;
3. Input optical thickness
   It is recommended to input the optical thickness, if there are several variants input all needed values pressing the button "Enter"
4. Input single scattering albedo
   It is recommended to input the single scattering albedo (SSA), if there are several variants input all needed values pressing the button "Enter". If only optical thickness is varied, input the same value of SSA several times;
5. Input asymmetry parameter
   It is recommended to input the phase function parameter, if there are several variants input all needed values pressing the button "Enter".
6. Input ground albedo
   It is recommended to input the ground (surface) albedo, if there are several variants input all needed values pressing the button "Enter"
7. Input solar zenith angles (degrees)
   It is recommended to input the zenith solar incident angle (degrees), if you'd like to study the dependence on solar angle it is necessary to repeat the program several times with the same optical parameters and varying solar angle.
8. Input viewing zenith angles (degrees)

It is recommended to input the viewing zenith angle (degrees) if intensity is calculated.

9. Input viewing azimuth angle (degrees)
   It is recommended to input the viewing azimuth angle relative to the Sun (degrees) if intensity is calculated.
10. If Fluxes put 1 if Intensity put 0

Input "1" if calculating fluxes and "0" if calculating intensity.

The result is output in the file "*asymp.dat*" that is in the same directory as the program "ASYMP.exe". The input values of the model and calculated radiative characteristics are in the resulting file.

11.7.3 Requirements to the Report

Compile a concise report reflecting the principal stages and the obtained results of the performed exercise in form of tables and plots (graphs).

Chapter 12
Calculating Solar Irradiance with Eddington Method

Abstract The formulas of Eddington approximation as a kind of two-stream methods are presented for calculating solar irradiance in the atmosphere. The optical model of clear atmosphere is proposed for the practice implementation.

12.1 Eddington Approximation

Return to the equations system (1.29), obtained in the Chap. 1.

$$\begin{aligned} \frac{dH(\tau,\mu_0)}{d\tau} &= -[1-\omega(\tau)]I(\tau,\mu_0) + \omega(\tau)F_0\exp(\frac{-\tau}{\mu_0}) \\ 3\frac{dK(\tau,\mu_0)}{d\tau} &= -[3-\omega(\tau)x_1(\tau)]H(\tau,\mu_0) + \omega(\tau)F_0\exp(\frac{-\tau}{\mu_0}) \end{aligned} \quad (12.1)$$

The vertically heterogeneous atmosphere is assumed, i.e. the single scattering albedo $\omega_0(\tau)$ and phase function parameter $g(\tau) = x_1(\tau)/3$ are functions of the optical thickness. Till now all transformations with the transfer equation are strict. The following approximation is done further:

$$\int_{-1}^{1} I(\tau,\mu)\mu^2 d\mu = \frac{1}{3}\int_{-1}^{1} I(\tau,\mu)d\mu. \quad (12.2)$$

That is the average value μ^2 is factor out from the integral sign. This relation is strict if the intensity obeys to the following dependencies on the viewing angle:

1. $I \neq I(\tau,\mu)$; – the constant;
2. $I(\tau,\mu) = a + b\mu$ – is the linear dependence;
3. $I(\tau,\mu) = I(\tau,\mu) + \Sigma I_i(\tau,\mu)\mu^{2i+1}$ – the polynomial dependence.

Assuming the boarder conditions as: $2H(0,\mu_0) = -I(0,\mu_0)$; $2H(\tau_0,\mu_0) = -I(\tau_0,\mu_0)$ provide also the equality: $3\,K(\tau,\mu_0) = I(\tau,\mu_0)$.

I. Melnikova et al., *Remote Sensing of the Environment and Radiation Transfer*, DOI 10.1007/978-3-642-14899-6_12,

The Eddington method is the kind of *two-stream approximations*. These approximations of the transfer theory are based on different analytical formulas approximately representing angle distribution of upward and downward intensity in the plane media. The substituting these formulas to the integro-differential transfer equation provides the system of differential equations for upward and downward irradiances:

$$\begin{aligned}\frac{dF^{\uparrow}(\tau,\mu_0)}{d\tau} &= \gamma_1 F^{\uparrow}(\tau,\mu_0) - \gamma_2 F^{\downarrow}(\tau,\mu_0) - F_0\omega\gamma_3\exp(-\frac{\tau}{\mu_0})\\ \frac{dF^{\downarrow}(\tau,\mu_0)}{d\tau} &= \gamma_2 F^{\uparrow}(\tau,\mu_0) - \gamma_1 F^{\downarrow}(\tau,\mu_0) + F_0\omega\gamma_4\exp(-\frac{\tau}{\mu_0})\end{aligned} \tag{12.3}$$

where $F^{\uparrow\downarrow}(\tau,\mu_0) = 2\pi\int\limits_0^1 I(\tau,\mu_0,\pm\mu)\mu d\mu$ are solar upward and downward irradiances in the atmosphere.

In the case of absorption absence or *conservative scattering* it is true: $\omega_0 = 1.0$, and the illumination of the surface with the albedo A (the transmitted irradiance at the atmosphere base) is expressed as:

$$\bar{F}^{\downarrow}(\tau_0,\mu_0) = \frac{4}{4+(3-x_1)(1-A)\tau_0}\left[\left(\frac{1}{2}+\frac{3}{4}\mu_0\right)+\left(\frac{1}{2}-\frac{3}{4}\mu_0\right)\exp\left(-\frac{\tau_0}{\mu_0}\right)\right] \tag{12.4}$$

In case of vertically homogeneous absorptive atmosphere the expressions for reflected and transmitted irradiances are:

$$\begin{aligned}F^{\uparrow}(0,\mu_0) &= C_1 + C_2 + D;\\ F^{\downarrow}(\tau_0,\mu_0) &= C_1\exp(-k\tau_0) + C_2\exp(k\tau_0) + D\exp(\frac{\tau_0}{\mu_0}) + \exp(\frac{\tau_0}{\mu_0});\end{aligned} \tag{12.5}$$

Expressions for constants in Eq. 12.5 are:

$$C_1 = \frac{\omega_0}{2}\frac{1}{1-k^2{\mu_0}^2}\frac{1}{\Delta}\cdot$$
$$\left\{[2+3\mu_0+(1-\omega)x_1(1+2\mu_0)]\exp(k\tau)(1+b)+[2-3\mu_0-(1-\omega)x_1(1-2\mu_0)]\exp(\frac{\tau_0}{\mu_0})(1-b)\right\};$$

$$C_2 = -\frac{\omega_0}{2}\frac{1}{1-k^2{\mu_0}^2}\frac{1}{\Delta}\cdot$$
$$\left\{[2+3\mu_0+(1-\omega_0)x_1(1+2\mu_0)]\exp(-k\tau)(1-b)+[2-3\mu_0-(1-\omega)x_1(1-2\mu_0)]\exp(-\frac{\tau_0}{\mu_0})(1+b)\right\};$$

$$D = -\frac{3+(1-\omega_0)x_1}{1-k^2{\mu_0}^2}\mu_0\frac{\omega_0}{2};$$

$$\Delta = \exp(k\tau)(1+b)^2 - \exp(-k\tau)(1-b)^2;$$

$$b = \frac{2k}{3-\omega_0 x_1}, \quad k^2 = (1-\omega_0)(3-\omega_0 x_1);$$

Eddington formulas are approximate solution of the transfer equation; they do not take into account strictly the angular dependence of the radiation field. Thus these formulas do not provide the high accuracy for intensity calculation. But the considered approach is a convenient for irradiance and radiative divergence calculations. The detailed analysis demonstrates that this approach is the most exact and optimal within wide ranges of atmospheric optical parameters. The method of the delta-Eddington better includes the scattering anisotropy with the phase function parameter. For solar zenith angles <75° the uncertainty of the approach is about 1–3% depending on the optical model.

Another form of delta-Eddington formulas:

Expressions for the plane albedo (reflected irradiance in relative units of F_0 at the atmosphere top) $F^{\uparrow}(0,\mu_0)$ and transmission (the transmitted irradiance in relative units of F_0 at the atmosphere base or the illumination of the surface) $F^{\downarrow}(\tau_0,\mu_0)$ are the solution of the equation system (12.1) with boundary conditions $F^{\uparrow}(\tau_0,\mu_0) = F^{\downarrow}(0,\mu_0) = 0$ (i.e. surface albedo is zero), where μ_0 is the cosine of zenith solar incident angle. The result is might be written as:

$$F^{\uparrow}(0,\mu_0) = m_6\left[(1-\kappa\mu_0)(a_2+\kappa\gamma_3)e^{\kappa\tau_0} - (1+\kappa\mu_0)(a_2-\kappa\gamma_3)e^{-\kappa\tau_0} - 2\kappa(\gamma_3 - a_2\mu_0)\exp(-\frac{\tau}{\mu_0})\right]$$

$$F^{\downarrow}(\tau_0,\mu_0) = [1 - m_6(m_1 - m_2 - m_3)]\exp(-\frac{\tau}{\mu_0}) \tag{12.6}$$

where the following notions are used:

$$m_1 = (1+\kappa\zeta)(a_1+\kappa\gamma_4)e^{\kappa\tau}, \quad m_2 = (1-\kappa\zeta)(a_1-\kappa\gamma_4)e^{-\kappa\tau},$$
$$m_3 = 2\kappa(\gamma_4 + a_1\psi)e^{\frac{\tau}{\zeta}}, \quad m_4 = (\kappa+\gamma_1)e^{\kappa\tau} + (\kappa-\gamma_1)e^{-\kappa\tau}, \tag{12.7}$$
$$m_5 = 1 - \kappa^2\zeta^2, \quad m_6 = \frac{\omega'}{m_4 m_5},$$

$$a_1 = \gamma_1\gamma_4 + \gamma_2\gamma_3, \quad a_2 = \gamma_1\gamma_3 + \gamma_2\gamma_4,$$
$$\kappa = \sqrt{\gamma_1^2 - \gamma_2^2} \tag{12.8}$$
$$\gamma_1 = \frac{1}{4}[7 - \omega'(4+3g')], \quad \gamma_2 = -\frac{1}{4}[1 - \omega'(4-3g')]$$

$$\gamma_3 = \frac{1}{4}(2 - 3g'\zeta), \quad \gamma_4 = 1 - \gamma_3, \quad 3g = x_1 \tag{12.9}$$

The above set of formulas is the realization of the Eddington method. The phase function considering leads to delta-Eddington method, where optical parameters ω' and τ', transform in according to expressions:

$$\omega' = \omega_0 \frac{1 - g^2}{1 - \omega_0 g^2}, \quad \tau_0 = \tau' = \tau(1 - \omega_0/g^2), \quad g' = \frac{g}{1+g}, \tag{12.10}$$

12.2 Considering the Surface Reflection

The taking into account the surface albedo $A > 0$ is done according to known relations

$$\begin{aligned} \bar{F}^{\uparrow}(0,\zeta) &= F^{\uparrow}(0,\mu_0) + A_s V(\tau')\bar{F}^{\downarrow}(\tau_0,\mu_0) \\ \bar{F}^{\downarrow}(\tau_0,\mu_0) &= \frac{F^{\downarrow}(\tau_0,\mu_0)}{1 - A_s A(\tau_0)} \end{aligned} \tag{12.11}$$

where $A(\tau = 0)$ and $V(\tau_0)$ are spherical albedo and the transmittance and defined by expressions:

$$\begin{aligned} A(\tau = 0) &= 2\int_0^1 F^{\uparrow}(0,\mu_0)\mu_0 d\mu_0 \\ V(\tau_0) &= 2\int_0^1 F^{\downarrow}(\tau_0,\mu_0)\mu_0 d\mu_0 \end{aligned} \tag{12.12}$$

It seems easy to calculate the result that after integrating. But integrals of Eq. 12.7 do not lead analytical expressions directly. For diffuse radiation the result was obtained for reflected radiation $A(0)$ and for transmittance $V(\tau_0)$ without considering the direct radiation:

$$\begin{aligned} A(0) &= \frac{\gamma_2(1 - e^{-2\kappa\tau_0})}{\kappa + \gamma_1 + (\kappa - \gamma_1)e^{-2\kappa\tau_0}} \\ V(\tau_0) &= \frac{2\kappa e^{-\kappa\tau_0}}{\kappa + \gamma_1 + (\kappa - \gamma_1)e^{-2\kappa\tau_0}}, \end{aligned} \tag{12.13}$$

The direct radiation for transmittance might be taken into account by addition the item in the expression for $V(\tau_0)$:

$$\int_0^1 \mu \exp\left(\frac{-\tau_0}{\mu}\right) d\mu = -\exp(-\tau_0) + \tau_0 E_1(\tau_0)$$

where $E_1(\tau_0)$ is exponential integral, expansions are true for it:
for small argument (optical thickness $\tau_0 < 1$):

$$E_1(\tau) = -0.5772 - \ln \tau_0 + \tau_0 - \frac{\tau_0^2}{2 \cdot 2!} + \frac{\tau_0^3}{3 \cdot 3!} + \ldots, \qquad (12.14)$$

and for big argument (optical thickness $\tau_0 > 1$):

$$E_1(\tau) = \frac{\exp(-\tau_0)}{\tau_0}\left(1 - \frac{1}{\tau_0} + \frac{2}{\tau_0^2} - \frac{6}{\tau_0^3} + \ldots\right).$$

Thus the direct radiation in the expression (12.14) is accomplished differently depends on the optical thickness. Point out that $E_1(\tau_0)$<0.01 for the optical thickness $\tau_0 > 4$ and the direct radiation might be omitted.

12.3 Calculation of Radiation Characteristics

The estimating the diffuse radiation rate in the transmitted irradiance might be important for certain problems. The expression for it is follows:

$$tt = \frac{\bar{F}^{\downarrow}(\tau_0, \mu_0) - \exp(\frac{\tau_0}{\mu_0})}{\bar{F}^{\downarrow}(\tau_0, \mu_0)} \qquad (12.15)$$

In some problems the ratio of the irradiance, reflected by the atmosphere at the top, to the transmitted one at the base (should not be confused with surface albedo, where the irradiance reflected by the surface at the atmosphere base is in the definition) might be useful.

$$rr = \frac{\bar{F}^{\uparrow}(\tau = 0, \mu_0)}{\bar{F}^{\downarrow}(\tau_0, \mu_0)}. \qquad (12.16)$$

It is not difficult to calculate *the direct and diffuse irradiance at the vertical surface* (e.g. the UV irradiance at standing person):

$$\bar{F}^{\downarrow}{}_{vert}(\tau_0, \mu_0) diffused = \frac{\bar{F}^{\downarrow}(\tau_0, \mu_0) - \exp(-\tau_0/\mu_0)}{2}$$
$$\bar{F}^{\downarrow}{}_{vert}(\tau_0, \mu_0) direct = \sqrt{1 - {\mu_0}^2} \exp(-\tau_0/\mu_0) \qquad (12.17)$$

The overall irradiance (diffused + direct) at the vertical surface, oriented to the Sun is the summation of Eq. 12.17:

$$\bar{F}^{\downarrow}{}_{vert}(\tau_0,\mu_0)Sum = \sqrt{1-\mu_0{}^2}\exp(-\tau_0/\mu_0) + 0.5 \times \left[\bar{F}^{\downarrow}(\tau_0,\mu_0) - \exp(-\tau_0/\mu_0)\right] \quad (12.18)$$

The radiative divergence is the important characteristic for the atmospheric radiation regime:

$$\bar{R}(\mu_0) = 1 - \bar{F}^{\downarrow}(\tau_0,\mu_0) - \bar{F}^{\uparrow}(0,\mu_0) \quad (12.19)$$

It is to remind that all above radiative characteristics are in relative units. For obtaining them in energetic units it is necessary to multiply to $F_0\mu_0$.

12.4 The Atmosphere Optical Model

It is necessary to input the optical model of the atmosphere (the set of optical parameters describing the media) for calculating radiative characteristics – the direct problem solution. Here we consider the simplest variant of the vertically homogeneous atmosphere. It has been shown that this approximation provides the irradiance uncertainty less than 10% for irradiance calculation. Assume the molecular atmosphere together with scattering and absorbing aerosols. The shortwave spectral range is considered. The following values are input:

- The optical thickness of the clear atmosphere $\tau_0 = \tau_{as} + \tau_{aa} + \tau_{ms} + \tau_{ma}$, where τ_{as} and τ_{ms} are optical thicknesses of the aerosol and molecular (Rayleigh) scattering, τ_{aa} and τ_{ma} are optical thicknesses of the aerosol and molecular absorption;
- The optical thickness of cloud $\tau_{cl} = \tau_{cs} + \tau_{ca}$ is the sum of optical thicknesses of the cloud scattering and absorption;
- The single scattering albedo (probability of the photon surviving while single interaction)
  for clear atmosphere $\omega_0 = (\tau_{as} + \tau_{ms})/\tau_0$;
  for cloudy atmosphere $\omega_0 = (\tau_{cs} + \tau_{as} + \tau_{ms})/(\tau_{cl} + \tau_0)$;
- The phase function asymmetry parameter $g = 0$–0.3 for the clear atmosphere and $g = 0.8$ for cloud;
- The surface albedo A_s.

Examples of numerical values are in the Tables 12.1 and 12.2. The molecular absorption is neglected in the shortwave region. It is possible to add to the total optical thickness the cloudy optical thickness $\tau_{cl} = 10, 20$ for the cloudy atmosphere and to recalculate the single scattering albedo in agreement with the above definition.

12.5 Practice 11

12.5.1 Objectives

- to calculate solar radiative characteristics for models of the clear and (or) cloudy atmosphere;
- to study dependences of irradiances and radiative divergence on optical parameters and solar zenith angle;
- to compare with values obtained for cloudy atmosphere with asymptotic method, or clear atmosphere – by single scattering approximation and by Monte-Carlo approach.

12.5.2 Sequential Steps of the Exercise Implementation

The following radiative characteristics are proposed to calculate for the set of values of solar zenith angle, optical thickness, single scattering albedo, phase function parameter, and surface albedo:

1. the reflected and transmitted irradiances. It is noted in the resulting file as "refl" and "trans"
2. the radiative divergence (div),
3. the rate of diffuse radiation in the transmitted irradiance (dif/transm)
4. the ratio of reflected to transmitted irradiance (refl/transm)
5. the direct irradiance to vertical surface (vertDirect)
6. The overall irradiance (diffused + direct) to vertical surface (vertSum)

Calculation is accomplished with the compute program "edding.exe" that is compiled in Borland 4 C++. After the program running the writing is output to the screen: "Program calculates radiative characteristics and diffusive part with D-Eddington"

The program "EDDING.exe" proposes the following dialog when questions are output at the screen sequentially.

1. Input number of wavelength (group tau-lam-g-A)
   - It is recommended to input the number of calculation variants;
2. Input number of solar incident angles
   - It is recommended to input the number of solar angles;
3. Input wavelengths
   - It is possible just numerating: 1,2,3. . . or input values of the wavelength. This parameter does not participate in calculation;

4. Input optical thickness
   - It is recommended to input one or several values of the optical thickness (depending on the answer to question 1) pressing the button "Enter" every time. The range of values is 0.1–2.0 for the clear atmosphere or 2–20 for cloudy atmosphere. Values from the Table 12.1 might be use;

   If the dependency on one parameter is studied (e.g. optical thickness) the set of several (e.g.10) parameter values is input and other parameters are repeated the same.
5. Input single scattering albedo (SSA)
   - It is recommended to input one or several values of the SSA (depending on the answer to question (1). The range of values is 0.7 in case of strong absorption in the clear atmosphere, till 0.999 weak absorption in the cloudy atmosphere, the value 1.0 for pure (conservative) scattering is better to change for 0.99999999, otherwise the overflow might be happen. If only optical thickness is varied, input the same value of SSA several times;
6. Input phase function asymmetry parameter – It is recommended to input the phase function parameter, if there are several variants input all needed values pressing the button "Enter". The range of values is 0 (isotropic scattering – possible to use for the clear and pure atmosphere) till 0.85 (extended phase function in clouds)
7. Input ground albedo
   - It is recommended to input one or several values of the ground albedo (depending on the answer to question (1). The range of values is 0, 0.02 (the absence of the surface reflection above sea surface) till 0.95 (strong reflection by fresh snow);
8. Input solar incident angles (degrees)
   - It is recommended to input one or several values of the solar incident angle (depending on the answer to question 2 in ranges: 0° (the Sun is in zenith) till 80° (the Sun is near the horizon).

Then the program accomplishes the calculation and output the result to the screen and to the resulting file "*eddaer.dat*", where are also input values of the optical model

Table 12.1 The optical model of the clear atmosphere

λ, μm	0.36	0.4	0.45	0.5	0.55	0.6	0.65	0.7	0.8	0.9
τ_{as}	0.379	0.316	0.285	0.264	0.250	0.237	0.224	0.213	0.201	0.190
τ_{aa}	0.04	0.04	0.04	0.04	0.04	0.04	0.04	0.04	0.04	0.04
τ_{ms}	0.564	0.364	0.223	0.145	0.098	0.072	0.052	0.039	0.023	0.014
τ_0	0.983	0.720	0.552	0.449	0.388	0.349	0.316	0.292	0.264	0.244
ω_0	0.959	0.944	0.927	0.911	0.897	0.885	0.873	0.863	0.849	0.836
g	0.3	0.3	0.3	0.3	0.3	0.3	0.3	0.3	0.3	0.3

Table 12.2 Ground albedo for three types of surfaces: (I) mowed grass, (II) sea surface and (III) fresh snow

A_I	0.03	0.035	0.045	0.056	0.070	0.087	0.110	0.136	0.193	0.225
A_{II}	0.05	0.06	0.07	0.07	0.05	0.04	0.03	0.03	0.03	0.03
A_{III}	0.83	0.85	0.87	0.90	0.90	0.90	0.90	0.89	0.87	0.86

Solar zenith angles are arcos $\mu_0 =$ from 0° till 75°

12.5.3 Requirements to the Report

Compile a concise report reflecting the principal stages and the obtained results of the performed exercise in form of tables and plots.

It is to plot dependences of irradiances and radiative divergence versus the optical model (optical thickness, single scattering albedo, phase function parameter, surface albedo) and solar zenith angle, and prepare the report.

Chapter 13
Monte-Carlo Method for the Solar Irradiance Calculation

Abstract The base of the Monte-Carlo method is considered for atmospheric optics application. The algorithm of calculating hemispherical fluxes and radiative divergence is discussed. The description of the practice is presented.

13.1 The Basic Idea of Monte-Carlo Method

The Monte-Carlo method (more strict name is the method of statistical modelling) is a most powerful method of the radiative transfer theory. It allows to solve the problems concerned the radiance calculation with taking into account spherical geometry, polarization, heterogeneity of the atmosphere and surface, etc. Here we are applying this method for solving rather simple (comparing with above-mentioned) problem of the solar irradiance calculation in the horizontally homogeneous and plane parallel atmosphere. The approach allows simulations of the physical processes of radiative transfer in the atmosphere, when it is not needed to attract a body of the transfer theory. The Monte-Carlo method main idea is the interpretation of photon–atmosphere interaction as the random process: the motion of light conditional particle called "photon", the computer simulation of the process, and the calculation of desired characteristics as a mathematical expectation of random numbers appearing during the simulation. It is to be mentioned that here the photon is a mathematical object (not physical particle) and it might be divided to parts in further consideration. Desired radiative characteristics (radiance and irradiance) are obtained as average values over a multiplicity of photon trajectories sequent simulated.

The simplest example is considered for clarity, namely the plane atmosphere with the optical thickness τ_0 is illuminated by incident solar flux F_0 and cosine of the incident angle is μ_0. The transmission function defining the solar energy extinction in the atmosphere according to Beer's law is $P = \exp(-\tau_0/\mu_0)$. The P is possible to treat as *the probability* of single photon passes throughout the atmosphere (actually, $0 \leq P \leq 1$). Consider N photons and simulate their motion throughout the atmosphere as follows: take the random number α from the totality

DOI 10.1007/978-3-642-14899-6_13,

of the random numbers uniformly distributed over the interval [0, 1], the individual for every photon, and if the inequality $\alpha \leq P$ is true, consider that the photon passes throughout the atmosphere, otherwise – it does not pass. Calculate number of all passed photons: $N(\tau_0,\mu_0)$. The incident flux at the atmosphere top $F_0\mu_0$ recalculating to one photon leads to the photon energy $F_0\mu_0/N$. The flux (irradiance) at the atmosphere base is derived after multiplying this energy to the number of photons passing throughout the atmosphere $F^{\downarrow}(\tau_0,\mu_0) = \frac{F_0\mu_0}{N}N(\tau_0,\mu_0)$.

Unlike other methods in the Monte-Carlo method, it is appropriate not to divide radiation to the direct, diffused and reflected from the surface.

Certainly the simple solution might be written without this consideration (explain why $\frac{F_0\mu_0}{N}N(\tau_0,\mu_0) \to F_0\mu_0 \exp(-\tau_0/\mu_0)$ while $N \to \infty$ as an exercise). However, it is convenient to obtain irradiance with real multiple scattering and absorptions (i.e. to solve the general radiative transfer problem) if three processes are successfully modelled: photon-atmosphere interaction (scattering and absorption), photon-surface interaction (reflection and absorption) and the photon free path.

13.2 Simulating Random Events and Values

Consider some mathematical definitions before considering the Monte-Carlo method.

For the statistical simulation on computer, it is necessary to reproduce a process that will play the role of random event. Special algorithms are elaborated for random number choice that is called *a random number digitizer* (RND) or *randomizer* and there are special computer programs for generating different sequences of random numbers including to programming languages. However, testing of such ready algorithms shows no convenient distribution parameters of sequences (the mean and dispersion) for scientific purposes. Hence improved algorithms of RND have been elaborated for the Monte-Carlo method. This choice plays the role of "blind chance" similar to roulette wheel or shuffling the cards. Just this analogue causes the name of the method.

The totality of the random numbers uniformly distributed over the interval [0, 1] is the base of the Monte-Carlo method. We are implying only these numbers using the term "the random number", specifying them by sign β, and *at every its appearance in the text we mean a new random number*.

Let the probability of a certain discrete random event be equal to P. Choose the random number and if $\beta \leq P$, then assume that the event has happened, in the opposite case assume that it has not happened. The grounds of this approach are evident: if the quantity of the simulating acts tends to the infinity then the ratio of the quantity of the simulating acts, when the event has happened, to the quantity of all acts is equal to the probability of the event i.e. to P due to the uniformity of the random numbers distribution. Simulating random values is needed aside from random events. Note that according the definition the probability value u within the interval $[a,u]$ is equal to $P(u) = \int\limits_a^u \rho(u')du'$ for simulating the continuous random value characterized with *the probability density* $\rho(u)$ within the interval $[a,b]$. The application of the above-mentioned approach for discrete random values to simulating continuous values u leads to the following equation:

$$\int_a^u \rho(u')du' = \beta. \tag{13.1}$$

The Eq. 13.1 is the equation relative to the integral upper limit for obtaining the value u from the random number β.

13.2.1 Simulating the Photon Free Path

As it has been mentioned above, the process of radiative transfer in the Monte-Carlo method is simulated as a photon motion. Coming to the atmosphere the photon is moving along a certain trajectory, which finishes either with its outgoing from the atmosphere or with its absorption in the atmosphere or at the surface. Let the photon is at the optical depth τ_1 with the cosine of angle of its motion direction μ. A free photon path is analogous to the transfer of solar direct radiation throughout the atmosphere. The probability to reach a certain optical depth τ_2 is defined by Beer's Law: $P(\tau_2) = \exp(-(\tau_1-\tau_2)/\mu_0)$. The opposite event is the interaction with the atmosphere before the level τ_2 is interested for the consideration and the probability is $1-P(\tau_2)$. The probability density of the value τ_2 distribution is $\frac{\partial}{\partial \tau_2}(1-P(\tau_2))$ according to the definition.

$$\rho(\tau_2) = \frac{1}{\mu}\exp(-(\tau_2-\tau_1)/\mu) \tag{13.2}$$

The simulating of the photon free path is obtained after substituting (13.2) to the Eq. 13.1

$$\tau_2 = \tau_1 - \mu\ln(1-\beta) \tag{13.3}$$

Note that the Eq. 13.3 is true for both photon motion downward (for $\mu > 0$ it is $\tau_2 > \tau_1$) and upward (for $\mu < 0$ it is $\tau_2 < \tau_1$)

13.2.2 Simulating Photon-Atmosphere Interaction

The single scattering albedo $\omega(\tau)$ is treated as the probability of scattering photon interacting with the atmosphere at the optical depthτ. This value is called also the probability of the photon surviving that is more illuminating for the Monte-Carlo method. Here the photon absorption is treated as the photon death (it is swallowed by molecules or aerosols). The probability of the photon scattering in the atmosphere is $\omega_0(\tau)$. Thus if $\beta \leq \omega_0(\tau)$, then the photon scattering is occurring in the opposite case the absorption is happening i.e. the end of the trajectory. Cosine of the scattering angle χ and azimuth of the scattering Ψ are to be obtained for a new photon direction after scattering. The phase function $x(\tau,\gamma)$ describes the probability also – it is the probability density of scattering to the angle γ. Then the Eq. 13.1 is solved for simulating the scattering angle. The phase function is input as look-up table with the linear interpolation that leads to the Eq. 13.1 transforming to the square equation.

But assuming the Henyey-Greenstein function (1.15) with the one parameter g meaning the scattering angle mean cosine is more transparent. After a certain transformation it is obtained

$$\frac{1}{2}\int_{-1}^{\chi} x(\chi', g(\tau))d\chi' = \beta \tag{13.4}$$

The integral in the Eq. 13.4 is explicitly calculated, and the formula for the model of scattering angle cosine is derived as (it is recommended to do corresponding transformations yourselves)

$$\chi = \frac{2\beta(1+g^2(\tau))(g(\tau)\beta+1-g(\tau)) - (1-g(\tau))^2}{(2g(\tau)\beta+1-g(\tau))^2} \tag{13.5}$$

There is a second coordinate – the scattering azimuth angle Ψ. But it is simply simulated as uniformly distributed value in the interval $[0,2\pi]$ because of considered phase functions are not function of azimuth:

$$\Psi = 2\pi\beta \tag{13.6}$$

Thus the geometry of the scattering is defined. It is necessary to describe the photon direction after interaction. Let the photon moves before the scattering with the zenith angle cosine μ_1 and azimuth angle φ_1, then it change the direction to the angle cosine χ and azimuth Ψ. It is necessary to find new coordinates (μ_2,φ_2). The problem is solved with spherical trigonometry formulas:

$$\mu_2 = \mu_1\chi - \sqrt{(1-\mu_1^2)(1-\chi^2)}\cos\Psi, \quad \cos(\varphi_2-\varphi_1) = \frac{\chi-\mu_1\mu_2}{\sqrt{(1-\mu_1^2)(1-\mu_2^2)}} \tag{13.7}$$

13.2.3 Simulating Photon-Surface Interaction

It is possible to attribute an evident meaning of the reflection probability to the albedo in the description of the interaction with the surface: the reflection occurs if

$\beta \leq A$ and the opposite case corresponds to the photon absorption by the surface and to the end of the photon trajectory. Due to the reflection of the orthotropic surface all possible directions of the photon are uniformly distributed and simulating of the new direction (μ_2,φ_2) after reflection yields

$$\eta_2 = -\cos(\pi\beta/2), \quad \varphi_2 = 2\pi\beta \tag{13.8}$$

13.3 Monte-Carlo Method General Algorithm

When assembled together above described acts of interaction the totality deals the general scheme of the algorithm:

1. The trajectory of the photon begins at the atmosphere top and his optical depth is $\tau = 0$, the initial angle cosine is $\mu = \mu_0$, and the azimuth $\varphi = 0$.
2. There are variants of the photon's fate after simulating the free path according to Eq. 13.3: if the photon reaches the surface after the path $\tau_2 \geq \tau_0$, the interaction with the surface is simulated; if the photon is still in the atmosphere $\tau_2 < \tau_0$, the interaction with the atmosphere is simulated.
3. Then the new direction of the photon is simulated according to Eqs. 13.8 or 13.6–13.7, if it does not be taken up, and the free path with taking into account the new direction is simulated with the Eq. 13.3 further. At this stage it is analyzed if the photon at the surface or still in the atmosphere.
4. If the new direction gives $\mu < 0$ (upward motion) the third possibility arises $\tau_2 < 0$: the photon leaves the atmosphere escaping to the space.

Thus the whole trajectory of the photon is simulated. The trajectory finish corresponds to the photon absorption at the surface or in the atmosphere and it escaping to the space. After the trajectory finish the following photon trajectory from the atmosphere top is simulated.

The above considered example approach for obtaining desired irradiance values is applied for counting photons. Let the downward $F^{\downarrow}(\tau)$ and upward $F^{\uparrow}(\tau)$ irradiances being found at the optical depth τ. Computer variables $N^{\downarrow}(\tau)$ and $N^{\uparrow}(\tau)$ called "counters" are assumed. In the beginning of simulation (before the first photon trajectory) they are zeroth. Further modelling of the photon free path Eq. 13.3 the cases of photon crossing the level τ is analyzed: Mathematically it means $\tau_1 \leq \tau \leq \tau_2$ (it is possible for $\mu > 0$) or или $\tau_1 \geq \tau \geq \tau_2$ (for $\mu < 0$). In the first case the photon crosses the level τ while downward moving, and the unity is added to the counter $N^{\downarrow}(\tau)$; in the second case the photon moves upward and the unity is added to the counter $N^{\uparrow}(\tau)$. These operations are named "writing to counters" (here writing the unity).

After simulating N trajectories desired irradiances are found following to formulas

$$F^{\downarrow}(\tau) = \frac{F_0\mu_0}{N} N^{\downarrow}(\tau) \quad \text{and} \quad F^{\uparrow}(\tau) = \frac{F_0\mu_0}{N} N^{\uparrow}(\tau). \tag{13.9}$$

The number of trajectories are theoretically to stream to infinity, usually it taken as tens or hundreds of thousands.

13.4 Modifications of Monte-Carlo Method

The basic limitation of the Monte-Carlo method is random uncertainty in results. It is possible to estimate simultaneously with irradiances calculation. The reason of this uncertainty is dispersion of values writing to counters at separate trajectories. For example one photon might cross the level τ, another photon might not because absorbed in the atmosphere. It is to consider as many as possible simulated acts at every trajectory for minimizing the dispersion. Another words it is to keep every photon avoiding its premature death. The first measure is refusing from breaking the trajectory while the photon absorption. Because here the photon is a mathematical object it is possible to divide it to parts. For example the photon part equal to the albedo A is reflected while interacting with the surface and the part $1–A$ is taken up by the surface. It is similar in the atmosphere with the single scattering albedo $\omega(\tau)$. The special variable the photon weight w is introduced for taken into account the photon surviving part. In the trajectory beginning it is $w = 1$. After every interaction in the atmosphere it is multiplied to $\omega(\tau)$ and after interaction with the surface – to the surface albedo A. Thus the weight decreases after every interaction. The type of the interaction (scattering or absorption) is not simulated because it is taken into account by recalculating the weight and trajectory continues with simulating the photon new direction. The current value of the photon weight w (not the unity) is written to counters.

The photon escaping from the atmosphere to the space finishes also the trajectory, while upward moving. Let the photon be at the optical depth τ_1. The probability of it escaping through the atmosphere top is $P = \exp(\tau_1/\mu)$ for $\mu < 0$ according with the Beers Law. But it is possible to consider that the part equal to P escapes, and the part equal to $1–P$ remains in the atmosphere and continue the trajectory. Hence, if $\mu < 0$ it is to write the escaping part wP to all counters of the upwelling irradiance with $\tau \leq \tau_1$, then before the simulating the free path to multiply the photon weight w to the value $1 - \exp(\tau_1/\mu)$. It is also to modify the free path simulating algorithm for the photon necessarily remains in the atmosphere. The normalizing of the probability density with the Eq. 13.2 is used for the probability (the integral from the density) is equal to the unity, when $\tau_2 = 0$. Another words the photon with the unity probability (necessarily) remains in the atmosphere. This demand leads to the relation:

$$\rho(\tau_2) = \frac{1}{\mu} \exp(-(\tau_2 - \tau_1)/\mu) \Bigg/ \int_{\tau_1}^{0} \frac{1}{\mu} \exp(-(\tau' - \tau_1)/\mu) d\tau' \tag{13.10}$$

That yields the following:

$$\rho(\tau_2) = \frac{1}{\mu}\frac{\exp(-(\tau_2 - \tau_1)/\mu)}{1 - \exp(\tau_1/\mu)} \tag{13.11}$$

Substituting the probability density (13.11) to the Eq. 13.1 results simulating the photon free path while upward moving ($\eta < 0$) without leaving the atmosphere. (Derive formula yourselves as an exercise)

$$\tau_2 = \tau_1 - \mu \ln(1 - \alpha(1 - \exp(\tau_1/\mu))) \tag{13.12}$$

With such approach the photon trajectory neither might nor finish at all and it is cut with taking into account the value of the weight. The trajectory simulating is terminated when the photon weight w is less than certain value fixed a priori.

Another useful modification of the Monte-Carlo method is the application of simulating without "escaping" to the procedure of writing to counters. Let the photon with a zenith angle cosine μ be at the optical depth τ_1. The probability to cross the optical depth τ ($\tau \geq \tau_1$ for $\mu > 0$, $\tau \leq \tau_1$ for $\mu < 0$) is $P = \exp(-(\tau - \tau_1)/\mu)$

And this value might be treated as the photon part reaching the depth τ and written to the corresponding counter. Thus values $w \exp(-(\tau - \tau_1)/\mu)$ are written to counters and this writing is done before recurrent simulating (not after) including the photon start from the top. There is no writing the photon free path according to Eqs. 13.3 and 13.12 to counters while the free path simulating.

Considered modification of the Monte-Carlo method significantly improves the algorithm and decrease the random uncertainty of calculated values.

13.4.1 *Azimuthal Isotropy of Irradiances*

It is clear from physical consideration that the irradiance does not depend on solar azimuth if the surface reflects orthotropic because the problem is axially symmetric and the photon coordinate "azimuth" φ is redundant. Let us consider its impact on calculated irradiances. Actually values written in counters do not depend on azimuth. Test the potential indirect dependence via the zenith angle cosine μ. Simulating the reflection from the surface according to Eq. 13.8 shows a random choice of the azimuth angle and the reflection angle cosine does not depend on azimuth. Similarly the zenith angle cosine μ_2 does not depend on initial azimuth φ_1 in agreement with Eqs. 13.6–13.8. Thus the coordinate "azimuth" appears unneeded, what simplified calculation.

13.5 Additional Possibilities of the Monte-Carlo Method

The algorithm has been considered for irradiance calculation. However other important radiative characteristics are possible to obtain without significant transformation. It is possible to count the photon parts taken up in the atmosphere and to write the value $w(1 - \omega(\tau))$ to the absorption counter. It is the way for the calculating the energy absorbed in the atmosphere that is very important for many problems in particular in climatic models. As it was mentioned in the Chap. 1 this energy named the radiative divergence in the atmosphere. The Monte-Carlo method is the only method providing the direct calculation of the radiative divergence.

It is possible to obtain certain fine characteristics of the process of radiative transfer that impossible with other approaches. In particular the ratio of the photon interacting with the atmosphere (or with the surface) is calculated and the distribution of desired values over this ratio. The arrays with index corresponding to the interaction ratio is to be used in spite of scalar counters. (Write the modification of the algorithm as an exercise). This distribution is important for a set of research problems e.g. it allows to estimate the exactness of single scattering approximation.

13.6 Practice 12

13.6.1 Objectives

The purpose of the practice is studying the dependence of irradiances transmitted by the atmosphere (illuminating the surface) and reflected from the atmosphere top and radiative divergence on atmosphere parameters and solar zenith angle. These values determine solar energy reaching the surface, escaping to the space and absorbing in the atmosphere.

For simplicity the homogeneous atmosphere model is used (i.e. optical parameters do not depend on the optical deepnessτ). Then only four parameters describe the atmosphere: the optical thickness τ_0, the single scattering albedo ω, the mean cosine of the scattering angle (phase function asymmetry parameter) g and the surface albedo A.

Take three the most interesting for study dependences:

1. The distribution desired values over interact ratio.
2. The dependence on solar zenith angle that point the variation of solar energy during the day time.
3. The dependence on the atmosphere optical thickness that demonstrates variation of desired values depending on atmospheric aerosols content.

Typical optical parameters of the atmosphere and surface are presented in the Chap. 10 (Eddington approximation).

However the input atmospheric model is not strictly linked with real values atmospheric parameters and they will be varied for better elucidating dependences of interest. The incident solar flux is assumed $F_0 = 1$.

13.6.2 Sequential Steps of the Exercise Implementation

Calculation is accomplished with the compute program "MMKFLUXS.exe" that realizes the algorithm considered above and is compiled in Borland 4 C++.

The file "MMKFLUXS.exe" proposes the following dialog when questions are output at the screen sequentially.

1. Input filename for results:

   It is recommended to input the file name where the result is output. The file is created for writing and while the program repeats the file is rewritten. It is open till the program running.

   Then the program accomplishes three stages:

   The first is the study of the distribution of considered characteristics: transmitted and reflected irradiance (or downward and upward flux) and radiative divergence (or flux divergences) over ratio of interaction (scattering and reflection). The sequential questions are output to the screen:

Input sun zenith angles (grad):

   It is recommend inputting one or two values of the solar zenith angle (degrees) Values of 45.0° and 80.0° are recommended.

   Attention! Several values divided by space in one line. For proposed example: 10.0 _45.0 "Enter"

Input optical depths:

   It is to input two or three values of the atmosphere optical thickness τ_0: corresponding to weak concentrations of atmospheric aerosols 0.1, average value 0.3 and heavy 0.7.

Input single scattering albedo values:

   Two or three values of the single scattering albedo are to be input. ω. For example: non absorbing aerosols 0.9999, weakly absorbing 0.9 and strongly absorbing 0.6.

Input phase function parameters:

   One or two values of the phase function parameter g are needed: for slightly extending phase function (fine fraction of aerosols) 0.2 and strongly extended (big aerosol particles) 0.75.

Input surface albedo:

   Two values of the surface albedo corresponding to snow 0.9, and water surface 0.05 are recommended.

Then the program calculates three radiative characteristics for all combinations of parameters including the contribution of every ratio of interaction (from zeroth till fifth and higher – six gradations). The contribution is calculated in percents from the total value. Results are output to the resulting file.

The second stage is studying the dependences of three calculated value on solar zenith angle. There are consequent questions:

Input optical depths:
Input single scattering albedo values:
Input phase function parameters:
Input surface albedo:

The answers the similar to above considered.

The program calculates desired characteristics ad output them to the resulting file as functions of the solar zenith angle for all input combinations.

The third stage is studying dependences of desired values on the optical thickness. There are sequent questions:

Input sun zenith angles (grad):
Input single scattering albedo values:
Input phase function parameters:
Input surface albedo:

Answers are similar to above considered. The program calculates desired characteristics as functions of all combinations of input parameters and output result to resulting file.

13.6.3 Requirements to the Report

Compile a concise report reflecting the principal stages and the obtained results of the performed exercise in form of tables and plots (graphs).

It is needed:

- Plotting obtained dependences (diagrams present the result in case of distribution over interaction ratio).
- Describing (and physically interpreting if possible) obtained dependences.

Chapter 14
Calculating Radiative Characteristics with the Single Scattering Approximation

Abstract The approximation of the single scattering is considered for case of homogeneous plane atmosphere. The surface reflection is taken into account. The description of the practice is presented.

14.1 Expressing the Radiative Intensity (Radiance) in Terms of the Source Function

Let us return to the transfer equation Eq. 1.25

$$\mu\frac{I(\tau,\mu,\mu_0,\varphi)}{d\tau}=-I(\tau,\mu,\mu_0,\varphi)+\frac{\omega_0(\tau)}{4\pi}\int\limits_0^{2\pi}d\varphi'\int\limits_{-1}^{1}x(\tau,\gamma)I(\tau,\mu',\mu_0,\varphi')d\mu'$$
$$+\frac{\omega_0(\tau)}{4\pi}F_0x(\tau,\gamma_0)e^{-\tau/\mu_0}\tag{14.1}$$

The diffuse radiation is treated as additional sources in the atmosphere, that gives the second and third items in the right side. Introduce *the source function*:

$$B(\tau,\mu,\mu_0,\varphi)=\frac{\omega(\tau)}{4\pi}\int\limits_0^{2\pi}d\varphi'\int\limits_0^{1}I(\tau,\mu',\mu_0,\varphi')x(\gamma,\tau)d\mu'+\frac{\omega(\tau)}{4\pi}F_0x(\gamma_0,\tau)\exp(-\tau/\mu_0)\tag{14.2}$$

Then the transfer equation develops to

$$\mu\frac{dI(\tau,\mu,\mu_0,\varphi)}{d\tau}=-I(\tau,\mu,\mu_0,\varphi)+B(\tau,\mu,\mu_0,\varphi)\tag{14.3}$$

I. Melnikova et al., *Remote Sensing of the Environment and Radiation Transfer*,
DOI 10.1007/978-3-642-14899-6_14, © Springer-Verlag Berlin Heidelberg 2012

The boarder conditions for the Eqs. 14.1 and 14.3 are

$$\begin{cases} I(\tau = 0, \mu, \mu_0, \varphi) = 0, \text{ for } \mu > 0 \\ I(\tau = \tau_0, \mu, \mu_0, \varphi) = 0, \text{ for } \mu < 0 \end{cases} \tag{14.4}$$

It corresponds to the absence of the incoming *diffuse* radiation at the atmosphere's top.

The Eq. 14.3 is an ordinary differential equation of the first order. (of type $\frac{dy}{dx} = a(x)y + b(x)$). It has a known general solution

$$(y(x) = y(x_0) \exp\left(\int_{x_0}^{x} a(x')dx'\right) + \int_{x_0}^{x} b(x') \exp\left(\int_{x'}^{x} a(x'')dx''\right) dx').$$

Applying it to the Eq. 14.3 ($x \equiv \tau$, $y \equiv I$, $a(x) \equiv -1/\mu$, $b(x) \equiv B/\mu$) by taking into account for the boarder conditions (14.4) ($x_0 \equiv 0$ for $\mu > 0$, $x_0 \equiv \tau_0$ for $\mu < 0$, for both cases $y(x_0) = 0$) yields:

$$\begin{cases} I(\tau, \mu, \mu_0, \varphi) = \dfrac{1}{\mu} \displaystyle\int_{0}^{\tau} B(\tau, \mu, \mu_0, \varphi) \exp\left(-\dfrac{\tau - \tau'}{\mu}\right) d\tau', \text{ for } \mu > 0 \\ I(\tau, \mu, \mu_0, \varphi) = -\dfrac{1}{\mu} \displaystyle\int_{\tau}^{\tau_0} B(\tau, \mu, \mu_0, \varphi) \exp\left(-\dfrac{\tau - \tau'}{\mu}\right) d\tau', \text{ for } \mu < 0 \end{cases} \tag{14.5}$$

Let's point out that expressions (14.5) are not the solution of the transfer equation (14.1) because the function $B(\tau, \mu_0, \mu, \varphi)$depends on desired intensity according to Eq. 14.2. Nevertheless they are convenient for consideration in certain cases.

14.2 The Approximation of the Single Scattering – The General Case

Consider the Eq. 14.2 for the source function. Two items have transparent physical sense. The first one expresses sources of *the multiple scattered light* in the atmosphere, i.e. scattered more than one time. The second item characterises sources of *the single scattered light* that arises by substituting the direct radiation to the transfer equation. The contribution of single scattered light to the total intensity significantly (several times) exceeds the contribution of multiple scattering in the clear Earth atmosphere. Thus it is possible to restrict *the single scattering approximation* and the first item in the expression for the source function to assume zeroth in many problems, where the high exactness is not needed. Then Eq. 14.5 give a real solution for the intensity calculation;

$$\begin{cases} I(\tau,\mu,\mu_0,\phi) = \dfrac{F_0}{4\pi\mu} \int\limits_0^{\tau} \omega(\tau')x(\gamma_0,\tau') \exp\left(-\dfrac{\tau'}{\mu_0} - \dfrac{\tau-\tau'}{\mu}\right) d\tau', \text{ for } \mu > 0 \\ I(\tau,\mu,\mu_0,\phi) = -\dfrac{F_0}{4\pi\mu} \int\limits_{\tau}^{\tau_0} \omega(\tau')x(\gamma_0,\tau') \exp\left(-\dfrac{\tau'}{\mu_0} - \dfrac{\tau-\tau'}{\mu}\right) d\tau', \text{ for } \mu < 0 \end{cases} \tag{14.6}$$

Equation 14.6 have a transparent physical sense and might be derived empirically. Let the solar radiation with initial flux F_0 scatter at the optical depth τ'. Then it decays from the atmosphere top to the level τ' according to Beer's law that corresponds to multiplying the flux to the value $\exp(-\tau'/\mu_0)$. Then the diffused intensity is expressed as $F_0\omega(\tau)'x(\gamma_0,\tau')/4\pi\ \exp(-\tau'/\mu_0)$. The radiation has a direction μ and passes the way τ–τ' till the level τ after the scattering event. Hence the radiation decays with multiplying to $\exp(-(\tau-\tau')/\mu)$ in accordance with the Beer's law and the result. And we consider that the scattering might occur at any levelτ' that needs integrating over the optical thickness. Finally the resulting expression coincides with Eq. 14.6.

14.3 The Single Scattering Approximation at Top and Base of the Homogeneous Atmosphere

In most cases the radiance at the atmosphere boarders (top and base) are interesting. Actually the solar radiation at the atmosphere base is important because it affects the biosphere. The reflected solar radiation at the atmosphere top is observed with many satellite instruments and it is important for remote sounding problems.

Thus the radiance at the base $I(\tau=\tau_0,\mu,\mu_0,\phi)$ for $\mu > 0$, and the radiance at the top $I(\tau=0,\mu,\mu_0,\phi)$ for $\mu < 0$ (Compare these values with the boarder conditions (14.4)). They are expressed with formulas (14.6), and integrating limits are from 0 to τ_0.

Introduce following notations for brevity:

$$I(\tau=\tau_0,\mu>0,\mu_0,\varphi) \equiv I_a^{\downarrow}(\mu,\mu_0,\varphi),\ I(\tau=0,\mu<0,\mu_0,\varphi) \equiv I_a^{\uparrow}(\mu,\mu_0,\varphi),$$

where index "a" points that it is the radiance of the radiation interacting only with the atmosphere.

The optically homogeneous atmosphere is assumed for simplifying i.e. optical properties and parameters do not depend on the optical depth τ: single scattering albedo ω and the phase function $x(\gamma)$ are not functions of τ. Then integrals in Eq. 14.6 easily derived and simple algebraic transformations (It is recommended as the exercise) yield:

$$
\begin{aligned}
I_a^{\downarrow}(\mu, \mu_0, \phi) &= \frac{F_0\mu_0}{4\pi}\omega x(\gamma_0)\frac{\exp(-\tau_0/\mu) - \exp(-\tau_0/\mu_0)}{\mu - \mu_0} \\
I_a^{\uparrow}(\mu, \mu_0, \phi) &= \frac{F_0\mu_0}{4}\pi\omega x(\gamma_0)\frac{1 - \exp(-\tau_0(1/\mu_0 - 1/\mu))}{\mu_0 - \mu}
\end{aligned}
\tag{14.7}
$$

14.4 The Surface Reflection

Equation 14.7 corresponds to single interaction between radiation and atmosphere. But the reflection from the surface contributes a significant part to the radiance. This reflected part at the atmosphere top might be much more than diffuse radiance (it is the reason why the surface is visible from the space). Hence taking into account that the surface reflection is necessary especially for interpretation of satellite images.

The total (diffused and reflected) radiance is written as:

$$
I^{\uparrow}(\mu, \mu_0, \varphi) = I_a^{\uparrow}(\mu, \mu_0, \varphi) + I_s^{\uparrow}(\mu, \mu_0, \varphi) \tag{14.8}
$$

where the value without index is the total one, and the index "s" points to the contribution of the surface reflection.

It is natural to consider only single interaction radiation-surface in range of the single scattering approximation. In addition it is assumed that after the reflection the radiation does not interact with the atmosphere because the opposite would be the second interaction. Then it is true according to the Beer's law:

$$
I_s^{\uparrow}(\mu, \mu_0, \varphi) = I_s(z = 0, \mu, \mu_0, \varphi)\exp(-\tau_0/\mu) \tag{14.9}
$$

where $I_s(z = 0, \mu, \mu_0, \varphi)$ is the intensity of the reflected radiation at the surface level $(z = 0)$.

The simplest model of the surface with the isotropic reflection is assumed. It means that reflected radiance does not depend on both the initial and reflected directions. Reflection is characterized by *the surface albedo A*. From the definition the albedo is the rate of reflected radiation:

$$
A = \frac{F^{\uparrow}}{F^{\downarrow}} \tag{14.10}
$$

where $F^{\downarrow}$ is the downward to the surface irradiance, $F^{\uparrow}$ is the upward from the surface (reflected) irradiance. These irradiances are expressed via radiance as the integral over the corresponding hemisphere as follows:

$$F^{\downarrow} = \int_0^{2\pi} d\varphi' \int_0^1 I(\mu', \varphi')\mu' d\mu'$$
$$F^{\uparrow} = -\int_0^{2\pi} d\varphi' \int_{-1}^{0} I(\mu', \varphi')\mu' d\mu' \quad (14.11)$$

It is enough calculating the irradiance $F^{\downarrow}$ for the albedo calculation and radiance $I_a^{\downarrow}(\mu, \mu_0, \varphi)$. However we must remember that $I_a^{\downarrow}(\mu, \mu_0, \varphi)$ is the diffuse radiance only and for the albedo calculation the direct radiance is to be added. The incident flux to the horizontal atmosphere top is $F_0\mu_0$. Transmitted through the atmosphere it decays according to Beer's law, hence at the base it is equal to $F_0\mu_0 \exp(-\tau_0/\mu_0)$ and the transmitted irradiance is:

$$F^{\downarrow}(\mu_0) = \int_0^{2\pi} d\varphi' \int_0^1 I_a^{\downarrow}(\mu', \mu_0, \varphi')\mu' d\mu' + F_0\mu_0 \exp(-\tau_0/\mu_0) \quad (14.12)$$

The integral in the Eq. 14.12 is calculated numerically.

(Explain why the radiance $I_a^{\downarrow}(\mu, \mu_0, \varphi)$ depends on azimuth φ).

The radiance reflected by the surface $I_s(z = 0, \mu, \mu_0, \varphi)$ does not depend on viewing angles μ and φ because the surface is orthotropic. Then $F^{\uparrow} = \pi I$ from the Eq. 14.11 and finally

$$I_s(z = 0, \mu_0) = \frac{A}{\pi} F^{\downarrow}(\mu_0) \quad (14.13)$$

14.5 The Single Scattering Approximation Algorithm

Thus the totality of Eqs. 14.8, 14.7, 14.9, 14.13, 14.12 provide the creation the algorithm for calculating the radiance at top and base of the homogeneous atmosphere with the orthotropic reflection at the base with assuming the single scattering approximation. Input data are:

1. atmospheric parameters: the optical thickness τ_0, single scattering albedo ω, phase function $x(\gamma)$, and surface albedo A;
2. geometric parameters: the cosine of the initial solar angle μ_0, cosine of viewing angle μ and the viewing azimuth φ;
3. the solar flux at the atmosphere top F_0.

All input parameters are scalars besides the phase function $x(\gamma)$. Generally it is a table, but here it is approximated by the Henyey-Greenstein function Eq. 1.16, that is defined with one parameter g and is possible to treat it as scalar.

$$x(\gamma) = \frac{1 - g^2}{(1 + g^2 - 2g\cos\gamma)^{3/2}} \tag{14.14}$$

Remember that the parameter g in Eq. 14.14 determines the function, varies in the range $0 \leq g < 1$, and coincides with the mean cosine of the scattering angle. The greater the value g the more extended the phase function is.

The important characteristic is obtained also: the transmitted irradiance at the base (14.12) shows the illumination of the surface.

14.6 Practice 13

14.6.1 *Objectives*

The purpose of the practice is studying dependences of the transmitted solar irradiance at the atmospheric base (14.12) and diffused solar radiances escaping from atmospheric boarders (14.8) on atmospheric parameters and geometry of the problem (viewing and solar directions). The following parameters are chosen:

1. The dependence of the transmitted solar irradiance at the atmospheric base on the solar zenith angle. This value characterizes variations of incoming to the surface solar energy during the day time.
2. The dependence of the transmitted solar irradiance at the atmospheric base on the optical thickness, because it demonstrates the impact of atmospheric aerosols on incoming solar energy to the surface. (The optical thickness increase called by aerosols concentration).
3. The dependence of diffused solar radiances escaping from atmospheric boarders on the optical thickness that characterizes the impact of atmospheric aerosols on the values, observed by satellite instruments.

14.6.2 *Applicability Ranges and the Input Optical Model*

The single scattering approximation is applicable, when the single scattered radiance is much more than the radiance multiple scattered. It is true while the small optical thickness τ_0 or the strong absorption (single scattering albedo ω is not close to the unity). The atmosphere optical thickness is the sum of molecular and aerosol components. The single scattering approximation is valid in spectral ranges, where the molecular contribution is neglected and only aerosols are taken into account, because the molecular optical thickness decreases dramatically with the wavelength (Rayleigh scattering). It is the visible and near infra-red ranges (different problems has their own restrictions).

Typical optical parameters are presented in Chap. 12.

But for studying dependences of interest optical parameters might be varied regardless of the real atmosphere values. The incident solar flux is assumed $F_0 = 1$.

14.6.3 Sequential Steps of the Exercise Implementation

The computer program "ODNOKRAT.exe" calculates desired characteristics. After running, the dialog is opened. The first question is

Input filename for results:

It is recommended to input name of the resulting file. Calculation results are output to this file, while program rerunning the file content is rewritten. It is open till the program is terminated.

The three stages are accomplished:

The first stage is studying the dependency of the transmitted irradiance (downward flux) on the solar incident angle (Sun zenith angle). There are sequent questions:

Input optical depths:

It is recommended to input two-three values of the optical thickness τ_0, e.g. corresponding to the low aerosol content 0.1, mean 0.3 and high 0.7.

Attention! Several values divided by space in one line. For proposed example: 0.1 _0.3_0.7 "Enter". Here the symbol "_" means space.

Input single scattering albedo values:

It is recommended to input two-three values of the single scattering albedo ω, e.g. nearly conservative aerosols 0.9999, weak absorption 0.9, and strong absorption 0.6.

Input phase function parameters:

It is recommended to input two-three values of the phase function asymmetry parameter g, e.g. values corresponded to near isotropic phase function 0.2 for fine aerosol particles, extended phase function 0.6 for 1–2 micron-sized particle, and strongly extended 0.75 for big aerosol particles.

The program calculates the transmitted irradiance as function of solar zenith angle for the set of all input parameters and output the result to fixed file.

The second stage is studying the transmitted irradiance dependence on the atmosphere optical thickness. There are questions:

Input solar zenith angles (grad):

Input two-three values of solar zenith angle (degrees). It is recommended values correspond to high solar position 45.0, mean value 65.0, and low Sun 80.0.

Input single scattering albedo values:

Input phase function parameters:

Answers are considered above.

The program calculates the transmitted irradiance as a function of the optical thickness for set of all input parameters and output to the fixed file.

The third stage is studying the diffused radiances (outgoing radiation intensity) dependences on the optical thickness. There are questions:

Input viewing directions - zenith angles and azimuths (grad):

The viewing zenith angle (*m* is the cosine), and azimuth *j* determine one viewing direction. Thus *the couple of values* are input, i.e. for two viewing directions input two couple (four values). Two directions are recommended: e.g. nadir direction: 180.0_0.0 are in most common use in remote sensing; and decline direction: 135.0_90.0. Take attention that for viewing of upward radiance the zenith angle exceeds 90°. Then there are questions:

Input sun zenith angles (grad):

Input single scattering albedo values:

Input phase function parameters:

Answers are considered above. It is recommended to restrict one-two input values for result amount will be reasonable.

Input surface albedo:

Input one-two surface albedo values, e.g. for snow 0.9, for vegetation and soils 0.25, for water surface 0.03.

Then the program calculates the outgoing radiances as function of the optical thickness for the set of all input parameters and output the result in fixed file.

The program is terminated.

14.6.4 Requirements to the Report

Compile a concise report reflecting the principal stages and the obtained results of the performed exercise in form of tables and plots (graphs). It is necessary:

- Plotting obtained dependences (diagrams present the result in case of distribution over interaction ratio).
- Describing (and physically interpreting if possible) obtained dependences.

Chapter 15
Analysis of the Reflection Anisotropy. Case Study: The Numerical Simulation of Waving Water Surface

Abstract The anisotropic reflection form waved water surface is considered. The characteristics of the reflected radiation: brightness coefficients and albedo are calculated. The description of the practice implementation is presented.

15.1 Types of Reflection from Natural Surfaces

Consideration of the optical properties of natural surfaces is the important part of many problems of the atmospheric optics and radiation transfer theory. It is especially essential for remote sensing of the atmosphere and surface.

Laws of interaction between radiation and surface is well known (the reflection angle is equal to the incident angle). However these laws have been formulated for ideal plane reflected surface. Real surfaces are irregular and (or) rough (grains of sand, soil clots, vegetation details). Thus the light reflects from every element in accordance with mentioned laws but the totality of elements (the surface) provides the multitudes of directions of reflected light (not the single direction as in the ideal case). This phenomenon is called *the diffuse reflection*. The reflection in according to the law "the reflection angle is equal to the incident one" is called here *the ideal mirror reflection*.

The diffuse reflection might be divided to the set of kinds. The simplest model *is the isotropic reflection*, when the reflected radiance is the same for all possible directions. It is this model that is used in many radiative transfer theory problems because of its simplicity. Corresponding surface is called *isotropic*. Only one parameter – *the surface albedo* is enough for characterizing the radiation-surface interaction. The albedo is the ratio of the reflected irradiance to the incident one. It is evident that the isotropy is the idealization. Fresh snow is the most near to the isotropic surface. If the reflected radiance is not the same to different directions it is called *the anisotropic reflection* and corresponding surface is called *the anisotropic surface*. *The quasi-isotropic* (i.e. the closest to ideal isotropic one, e.g. the fresh snow), *the quasi-mirror* (the closest to ideal mirror surface, e.g. the water surface) and *the inversely reflected* are possible to set off.

I. Melnikova et al., *Remote Sensing of the Environment and Radiation Transfer*, DOI 10.1007/978-3-642-14899-6_15, © Springer-Verlag Berlin Heidelberg 2012

The water surface is the mostly anisotropic of the reflections. It is very close to the plane i.e. ideally mirror, however waves disturb the idealization and transform the reflection from water surface to anisotropic diffuse (quasi-mirror). A correct consideration of the water surface is extremely important in many problems because it takes about 75% of the global Earth surface.

15.2 Statistical Simulation of the Waving Water Surface

The following approach is used for mathematical modelling optical properties of anisotropic surfaces. The surface is presented as the totality of elementary ideal planes. Reflection from every plane is simulated in accordance with the laws of ideal mirror reflection. The final reflected radiance is a sum over all planes.

Two types of models of water surfaces are possible for inclusion waves: dynamical and statistical. Dynamical models input the planes position as the time function. It corresponds to the real dynamics of reflection – in different time moments differently oriented planes reflect the light (the glares effect leads to twinkling when looking to the water). But the averaging picture is the most interesting in many cases. There is the photographing of the water surface from satellite or airplane board, the reflected radiance observation for example. The mentioned dynamics are averaged because reflected light from many planes are thrown to an instrument and separate glares are not distinguished. The most important case of the surface statistical model is used, where the position of separate planes is characterized by the function of probability density.

For waving water model it is the Cox-Munk function. In standard optical model of the atmosphere (e.g. Fig. 1.7) the orientation of the elementary plane characterises by unite vector normal to the plane (ϑ_n,φ_n), where ϑ_n is the nadir angle, φ_n is the azimuth. It is more convenient to use nadir angles (not zenith) for normals for they vary in the interval [0,π/2]. The nadir angle is zero for vector direction from nadir to zenith (upward perpendicularly to surface). The Cox-Munk function specifies the probability density for the normal to water surface and looks as follows:

$$P(\vartheta_n,\varphi_n) = \frac{\exp(-(x^2+y^2)/2)}{2\pi\sigma_x\sigma_y}\left(1 - \frac{1}{2}c_{21}(x^2-1)y - \frac{1}{6}c_{03}(y^3-3y) + 0.017(x^4-6x^2+3) + 0.03(x^2-1)(y^2-1) + 0.01Y\right)$$

$$\begin{aligned} &z_x = \sin(\varphi_n-\varphi_v)tg\vartheta_n, \quad \sigma_x = \sqrt{0.003+1.92\cdot 10^{-3}v}, \quad x = z_x/\sigma_x,\\ &z_y = \cos(\varphi_n-\varphi_v)tg\vartheta_n, \quad \sigma_y = \sqrt{3.16\cdot 10^{-3}v}, \quad y = z_y/\sigma_y,\\ &c_{21} = 0.01-0.0086v, \quad c_{03} = 0.04-0.033v, \quad Y = (y^4-6y^2+3) \end{aligned} \tag{15.1}$$

Parameters: wind velocity v [m/s] near surface and wind direction azimuth φ_v, are included in the Eq. 15.1. The Cox-Munk function (15.1) has been obtained with approximating measured declinations of sea waves. It seems a somewhat unwieldy, however it is simple because it is the two-dimensional normal distribution (the exponential) with certain corrections for taking into account real observations (items multiplied to the exponent). It is clear that in computer realization of the algorithm the unwieldiness of the Eq. 15.1 is not a problem.

Let the incident radiation with the initial zenith angle ϑ_1 and azimuth φ_1 fall to the surface. Then desired reflected radiance has the direction with the nadir angle ϑ_2 and azimuth φ_2 is found after three operations:

1. determining the elementary plane orientation (ϑ_n,φ_n) which reflects the light from the initial direction (ϑ_1,φ_1) to the direction (ϑ_2,φ_2);
2. calculating the reflection coefficient for this initial angle;
3. multiplying it to the probability density (15.1), incident radiance, and normalizing multipliers.

15.3 Laws of the Ideal Mirror Reflection

Consider the physical problem – calculating the reflection coefficient. Let the radiation fall to the ideal plane boarder of two media. Then the interaction between the radiation and surface is described by the following laws:

1. Incident radiation divided to two parts: reflected with direction from the surface to the first medium and the refracted with direction from the surface to the second medium.
2. Frequencies of the incident, reflected, and refracted radiation is equal.
3. Vectors of the incident, reflected, and refracted radiation and the normal to the surface are in one plane.
4. The reflected angle is equal to the incident angle.
5. The law determining the refraction angle (remember it as an exercise).

Note that these laws have been firstly obtained experimentally but they might be proved strictly theoretically. The same theory ascertains qualified relations between incident, reflected, and refracted radiances, and provides the expression for reflection coefficient as a direct consequence of general *Fresnel's formulas* looks as:

$$r(\beta) = \frac{1}{2}\left(\left(\frac{n^2\cos\beta - \sqrt{n^2 - 1 + \cos^2\beta}}{n^2\cos\beta + \sqrt{n^2 - 1 + \cos^2\beta}}\right)^2 + \left(\frac{\cos\beta - \sqrt{n^2 - 1 + \cos^2\beta}}{\cos\beta + \sqrt{n^2 - 1 + \cos^2\beta}}\right)^2\right) \tag{15.2}$$

where β is the incident radiation angle, n is the refraction coefficient of the surface matter. Here the radiation falls from the air, which refraction coefficient is equal to one unit. Then multiply the incident radiance to the expression (15.2) to obtain the reflected radiance.

15.4 Determining the Orientation of Elementary Plane

Directions of falling, (θ_1,φ_1) and reflecting (θ_2,φ_2) are prescribed. There's needed to find the normal to the elementary reflecting plane (θ_n,φ_n). It is more convenient to change to Cartesian coordinates $x_1 = \sin\vartheta_1 \sin\varphi_1$, $y_1 = \sin\vartheta_1 \cos\varphi_1$, $z_1 = \cos\vartheta_1$ that correspond to direction (θ_1,φ_1).

The angle between two directions (θ_1,φ_1) and (θ_n,φ_n) is the scalar product of vectors (it is recommended to do all transformation for exercise)

$$\cos\beta = \cos\vartheta_1 \cos\vartheta_n + \sin\vartheta_1 \sin\vartheta_n \cos(\varphi_1 - \varphi_n) \tag{15.3}$$

The angle β is the same as in Eq. 15.2.

The law of the incident and reflected angles equality yields the first equation for obtaining the desired direction:

$$\cos\vartheta_2 \cos\vartheta_n + \sin\vartheta_2 \sin\vartheta_n \cos(\varphi_2 - \varphi_n) = \cos\vartheta_1 \cos\vartheta_n + \sin\vartheta_1 \sin\vartheta_n \cos(\varphi_1 - \varphi_n) \tag{15.4}$$

The second equation is derived from the alignment of vectors and normal in one plane. It deals to the equality of the coordinate's determinant $\begin{vmatrix} x_n & y_n & z_n \\ x_1 & y_1 & z_1 \\ x_2 & y_2 & z_2 \end{vmatrix} = 0$ to zero that provides:

$$\begin{aligned} &\sin\vartheta_n \sin\varphi_n (\sin\vartheta_1 \cos\varphi_1 \cos\vartheta_2 - \cos\vartheta_1 \sin\vartheta_2 \cos\varphi_2) \\ &\quad + \sin\vartheta_n \cos\varphi_n (\cos\vartheta_1 \sin\vartheta_2 \sin\varphi_2 - \sin\vartheta_1 \sin\varphi_1 \cos\vartheta_2) \\ &\quad + \cos\vartheta_n \sin\vartheta_1 \sin\vartheta_2 \sin(\varphi_1 - \varphi_2) = 0 \end{aligned} \tag{15.5}$$

It is easy to express the $\mathrm{tg}\theta_n$ from the first Eq. 15.4, namely:

$$tg\vartheta_n = \frac{\cos\vartheta_2 - \cos\vartheta_1}{\sin\vartheta_1 \cos(\varphi_1 - \varphi_n) - \sin\vartheta_2 \cos(\varphi_2 - \varphi_n)} \tag{15.6}$$

In similar manner it is possible to go to tangent by dividing the second form Eq. 15.5 by $\cos\theta_n$, and the result after substituting the Eq. 15.6 to the Eq. 15.5 is the expression for the azimuth φ_n

$$tg\varphi_n = \frac{u}{w},$$

$$\begin{aligned} u &= (\cos\vartheta_1 - \cos\vartheta_2)(\cos\vartheta_1 \sin\vartheta_2 \sin\varphi_2 - \sin\vartheta_1 \cos\vartheta_2 \sin\varphi_1) \\ &\quad - \sin\vartheta_1 \sin\vartheta_2 \sin(\varphi_1 - \varphi_2)(\sin\vartheta_1 \cos\varphi_1 - \sin\vartheta_2 \cos\varphi_2) \\ w &= (\cos\vartheta_2 - \cos\vartheta_1)(\sin\vartheta_1 \cos\vartheta_2 \cos\varphi_1 - \cos\vartheta_1 \sin\vartheta_2 \cos\varphi_2) \\ &\quad + \sin\vartheta_1 \sin\vartheta_2 \sin(\varphi_1 - \varphi_2)(\sin\vartheta_1 \sin\varphi_1 - \sin\vartheta_2 \sin\varphi_2), \end{aligned} \tag{15.7}$$

Thus, formulas (15.7) and (15.6) solve the problem and found the direction of the normal vector to the elementary plane.

15.5 The Spectral Brightness Coefficient and the Albedo of the Waving Surface

Obtained above relations provide reflected radiance from the water surface if there is the incident radiance. It is more convenient to turn to reflecting properties of surface with considering the ratio of incident and reflected radiation (Kolmogorov and Fomin 1989). *The spectral brightness coefficient* is the characteristic of reflecting surface, $\rho(\vartheta_2, \varphi_2, \vartheta_1, \varphi_1)$, which is defined by the relation:

$$I(\vartheta_2, \varphi_2) = \frac{1}{\pi}\rho(\vartheta_2, \varphi_2, \vartheta_1, \varphi_1) I_0 \cos\vartheta_1 \tag{15.8}$$

where I_0 the incident, $I(\vartheta_2, \varphi_2)$is the reflected radiance. The sense of the spectral brightness coefficient is the ratio of the reflected intensity to the incident flux (irradiance) $I_0 \cos\vartheta_1$. The multiplier $1/\pi$ arises from the law of conservation of energy because the reflected energy (the integral of the function $I(\vartheta_2, \varphi_2)\cos\vartheta_2$ over the hemisphere) for isotropic surface (the reflected radiance does not depend on the direction) and absolutely white (reflects all incident radiation) has to be equal to the incident energy.

It is evident from the Eq. 15.8 that the spectral brightness coefficient of elementary plane is $\pi r(\beta)/\cos\beta$. The same value is possible to attribute to the total surface with specifying the density of probability of the needed plane appearance. Hence:

$$\rho(\vartheta_1, \varphi_1, \vartheta_2, \varphi_2) = \frac{\pi}{\cos\beta} P(\vartheta_n, \varphi_n) r(\beta), \tag{15.9}$$

where the spectral brightness coefficient dependence on the initial and reflected directions is governed by Eqs. 15.7, 15.6, 15.1, 15.3, and the reflection coefficient $r(\beta)$ is calculated with Eq. 15.2. There is an important property of the spectral brightness coefficient *symmetry*:

$$\rho(\vartheta_1, \varphi_1, \vartheta_2, \varphi_2) = \rho(\vartheta_2, \varphi_2, \vartheta_1, \varphi_1), \tag{15.10}$$

that corresponds physically to the reversibility of optical phenomena (reflective properties of the medium do not vary when the source and detector exchanged by places).

There are not necessary so detailed characteristic as the spectral brightness coefficient in many problems and only the ratio of reflected to incident energy. This characteristic, i.e. the rate of reflected radiation is called *the surface albedo*. It should be stressed that the albedo is characteristic of any surface not only isotropic. Just it is enough for isotropic surface and enough for anisotropic. The albedo is calculated as an integral of radiances over hemisphere. The following relation is the result

$$A(\vartheta_1, \varphi_1) = \frac{1}{\pi} \int_0^{2\pi} d\varphi_2 \int_0^{\pi/2} \rho(\vartheta_1, \varphi_1, \vartheta_2, \varphi_2) \cos\vartheta_2 \sin\vartheta_2 d\vartheta_2 \tag{15.11}$$

or, with taking into account for Eq. 15.9

$$A(\vartheta_1, \varphi_1) = \int_0^{2\pi} d\varphi_2 \int_0^{\pi/2} \frac{r(\beta)}{\cos\beta} P(\vartheta_n, \varphi_n) \cos\vartheta_2 \sin\vartheta_2 d\vartheta_2 \tag{15.12}$$

The anisotropic surface albedo depends in general on incident direction.

15.6 Practice 14

15.6.1 Objectives

The purpose of this practice is studying the dependence of spectral brightness coefficients and albedo of waving water surface on the incident direction, and near-surface wind velocity and azimuth.

The spectral range is visible interval (the value $n = 1.333$ is assumed for the water refraction index). Thus the considered model describes viewing effects of the solar illumination of the sea, which is studied in the practice. Note that our model does not correspond to reality because of the sky diffuse radiation and sea froth. But general effects of the radiation reflection from water surface are adequately described.

Two dependences are chosen for study as the most interesting:

1. The dependences of spectral brightness coefficient and albedo on solar elevation.
2. The dependence of Spectral Brightness Coefficient (SBC) and albedo on the wind velocity.

15.6.2 Sequential Steps of the Exercise Implementation

The computer program "SEEREFL.exe" realizes the above algorithm and calculates desired values. The dialogue on the screen is allowed.

The first question is:

Input filename for results:

It is necessary to input the file name for results output. The file is rewritten when the program is rerun and is open till the program terminates. Then two stages are accomplished.

The first stage is studying dependences of the spectral brightness coefficient and albedo on solar zenith angle. The consequent questions are output on the screen

Input wind speed (m/s):

It is recommended to input two-three values: the weak wind 2.0, moderate wind 7.5, and strong wind 15.0. Values are real numbers and are input with decimal point.

Attention! Several values divided by space in one line. For proposed example: 2.0 _7.5_15.0 "Enter"

Input wind azimuth (grad):

It is recommended to input one-two values of the azimuth: 0.0 и 90.0. The azimuth is reckoned from the Sun direction.

The program calculates desired values as functions of solar zenith angle for all set of input parameters at the definite network of viewing directions. The viewing direction is characterized with nadir angle and azimuth from the Sun direction. Result is formatted as a two-dimensional table (zenith angle and SBC) sequent output for all viewing directions and as three-dimensional tables (zenith angle, azimuth, and SBC) for all solar zenith angles. The latter provides two-dimensional plots as isometric lines.

The second stage is the study of the SBC and albedo dependence on the wind velocity.

The consequent questions are output on the screen

Input sun zenith angles (grad):

It is recommended two-three values of the solar zenith angle (in degrees) corresponding to high solar position 45.0, mean value 65.0, and low Sun 80.0.

Input wind azimuth (grad):

Answer is considered above.

The program calculates desired values as function of the wind velocity for all sets of input parameters and output to result file.

The program is terminated.

Finally you need to:

- Plot obtained dependences (diagrams present the result in case of distribution over interaction ratio).
- Describe (and physically interpret if possible) obtained dependences.

The SBC is to be presented as three-dimensional plots and find two effects: The first is the displacement of the brightness maximum at the solar path (for the zeroth azimuth) with variations of solar zenith angle.

15.6.3 Requirements to the Report

Compile a concise report reflecting the principal stages and the obtained results of the performed exercise in form of tables and plots (graphs). In the report answer to the question: why and how the maximum place is changed with the Sun moving from zenith to horizon? The second effect is the variation of the width of the solar path with the wind velocity changes. Is it possible to use this dependence for remote estimation the wind velocity?

Chapter 16
Quantification and Analysis of the Spectral Composition of Subsurface Solar Radiation Diffuse Reflectance in Cases of Deep and Shallow Water Bodies

Abstract Practice 15 in Chapter 16 is preceded by a concise theoretical description of light transfer in natural media, as well as inherent and apparent optical properties permitting to quantify the optical impact on the sunlight downward and upward propagation through the water column containing coexisting absorbing, scattering and fluorescing agents. A number of tables are provided to be used for numerically assess the above inherent and apparent optical properties as well as the water constituents optical properties that affect prevailingly the water colour variations in optically complex either inland marine coastal waters (so called case II waters as opposed to optically "simple" offshore oceanic waters coined case I waters). Also, the bottom optical properties are exemplified in form of tables permitting to take into account the optical bottom impact of the light emerging from below the water surface.

Based on these supplementary materials, the spectral values of subsurface diffuse reflectance, R(-0, l) can be simulated for the options of input parameters suggested in the tables. Qualitative analyses of simulated sets of the spectral envelope of R(-0, l) suggested to be conducted at the finalizing stage of Practice 15 performance is intended to give an insight into the composition and radiometric intensity of the light emerging from the surveyed water body and eventually captured by a remote sensor. This step is a predecessor to attaining the main goal of water remote sensing, viz, determination of water quality parameters.

16.1 Concise Theory

It is well-known that the World's Oceans play the key role in controlling both global cycles of a wide number of chemicals and their compounds and the bio-productional balance in the land-atmosphere system. In terms of biogeochemical cycles, the ocean presents a gigantic reservoir of nutrients. The salient chains of this reservoir functioning are the exchange of matter between lower and upper ocean layers as well as between the ocean and the atmosphere. Due to their gigantic

I. Melnikova et al., *Remote Sensing of the Environment and Radiation Transfer*, DOI 10.1007/978-3-642-14899-6_16, © Springer-Verlag Berlin Heidelberg 2012

thermal inertia, and at the same time active participation in heat transport, generation of cloud cover and a significant contribution to global carbon cycling, the World's oceans play the key role in the formation and dynamics of global climate.

Fully acknowledging the importance of the ecological state of open/pelagic regions of the oceans and associated seas, the importance of the ecological state of coastal marine and lacustrine zones should not be overlooked with regard to the overall status of the biosphere on the planet as well as in light of the fact that of a variety of sectors of economy in many countries are dependant totally upon the ecological state of such "marginal" aquatic environments. Such peripheral water zones are pivotal being the richest sources of food (marine food, in the first place), industrial raw materials/feedstock, as well as sources/resources of water for the benefits of economy and population.

However, due to significant spatial extension of aquatic bodies and, as a rule, remarkable dynamics of inherent biological processes, the traditional shipborne water sampling measurements are incapable to provide the space and time resolution required for adequate monitoring of the ecological state of such dynamic hydrological and biogeochemical environments. In this sense, remote sensing means/approaches provide most appropriate facilities, especially if they are mounted on aircraft or satellite platforms.

This naturally explains that a number of international and national organizations and agencies have launched/deployed wide-scale scientific research programs aimed at studying in depth the physical, chemical and biological processes (first and foremost, driven by anthropogenic forces) in the "Earth-Atmosphere" system. Importantly, such programs imply a wide use of remote sensing means of observation (predominantly constituting the payloads of environmental satellite platforms) to provide on a routine basis of quantitative assessments of key parameters characterizing the dynamics of on going changes.

By definition, remote sensing means provide data through indirect measurements. Most often, they operate with electromagnetic waves as information carriers. The signal adopted by the remote sensor is then analyzed by means of dedicated algorithms with a result of yielding the sought-for parameter.

The development of such algorithms is based on the knowledge and formal description (by various methodologies) of the processes of electromagnetic radiation transfer through both the object of investigation as well as the media intervening between the study object and the remote sensor.

When sounding aquatic media, it is reasonable to employ the range of the electromagnetic spectrum, which is less absorbed/attenuated by the target water column (i.e. the spectral range in which the water is most transparent). Such a range is confined between ~400 and ~ 700 nm, i.e. encapsulated in the visible part of the electromagnetic spectrum.

The Practices Nos. 15–17 are dedicated to studying optical properties of the water column in the visible with the application of such an apparent hydro-optical characteristic as the coefficient of light diffuse reflection.

This characteristic is a convolution of (a) absorbing and scattering properties of the water medium, and (b) geometry of propagation of the light flux beneath the water surface.

In the three Practices below, the specific features of sun light transfer in natural waters is investigated by means of numerical simulations followed by subsequent analysis of emerged changes in the spectral composition of the coefficient of diffuse reflection of sunlight immediately beneath the water surface, $R(-0,\lambda)$.

In Practice 15 the spectral variability in $R(-0,\lambda)$ is investigated as a function of concentration of color producing agents (CPAs) inherent in natural waters. By CPAs we assume the substances whose participation in the processes of sunlight absorption, scattering and Raman inelastic scattering emission effects control the intensity and spectral composition of light leaving the water-atmosphere interface. Such group of water constituents incorporates, inter alia, phytoplankton (whose cells encapsulate chlorophyll), mineral suspended matter and colored dissolved organics. As it will be stressed below, the water leaving radiative signal is also controlled by the bottom depth and its albedo.

Within the framework of the present Practices collection, the students are invited to familiarize themselves with the specific processes controlling the formation of water color characteristics (Practice 16) as well as with one of the methods of retrieval of CPA concentrations to the effect of remote sensing of natural water bodies (Practice 17).

A hydro-optical model developed for Lake Ladoga will be offered for effecting of the aforementioned exercises. The model is a set of tabulated spectral values of specific coefficients (cross-sections) of absorption and backscattering of the three CPAs, viz. phytoplankton chlorophyll, suspended mineral particulates and dissolved organic matter. This choice of CPAs is dictated by the typical/predominant hydro-optical constituents in Lake Ladoga. Through numerous investigations, it was shown that the suggested hydro-optical model of Lake Ladoga proves to be adequate for a wide range of hydro-optical conditions, and thus could be considered as confidently representative of waters at mid latitudes. The present set of exercises leaves beyond scope the pelagic marine waters as most hydro-optically simple, and thus not presenting some particular interest in light of studying the formation of water leaving radiative signal in the visible.

In addition to such mechanisms of interaction of solar light with the aquatic medium as absorption and elastic scattering, which participate in forming the radiance signal coming up from beneath the water surface in the case of all types of natural waters, in inland water bodies as well as marine coastal shallow zones, a significant impact on the spectral composition of water leaving radiance signal can be generated by such internal sources of radiation as phytoplankton fluorescence and dissolved organic matter (due to their typically high content). The other subsurface source of optical/radiative influence on the emerging light in such environments is constituted by the optical influence of the bottom. This in turn, can appreciably affect the accuracy of retrieval of water quality parameters.

When conducting remote sensing, it is mandatory to also consider the optical properties of the wind-roughened water surface as these characteristics/parameters determine the proportion of the desired signal in the total radiant signal captured by the remote sensor. However, within the framework of these teaching labs, for simplicity reasons we'll confine ourselves to the case of calm water surface.

The standard bio-optical water quality retrieval algorithms developed so far prove to be untenable unless they are employed strictly to off-coastal/ open ocean waters (so called case I waters). This is due to hydro-optically complexity of composition of inland and marine coastal waters (so called case 2 waters), which is further accentuated by the multitude of mechanisms of light interactions with the aquatic medium and pathways of formation of the water leaving radiant signal. Collectively, these factors render simplistic/case I water algorithms inadequate as they are traditionally based on band-ratio parameterizations operating with light signals in the blue and green spectral regions.

That is why when sounding case 2 waters, more accurate retrieval results can be attained employing more sophisticated mathematical approaches, among which is the Levenberg-Marquardt method of multivariate optimization. This procedure is based on minimization of squared sums of residuals between the measured and simulated/theoretical water volume diffuse reflectance. Practice No.10 is dedicated to this method and its application

Because all three Practice 15–17 are thematically closely interrelated, it is recommended to study the basic theoretical part of Chap. 16 before preparing to Practice 15–17. Besides, there are tables (Tables 16.1–16.4), which are intended as illustrations since the data incorporated in these tables can be automatically drawn

Table 16.1 Spectral water absorption and scattering coefficients and cross sections of absorption and backscattering of the main CPAs in the spectral region 410–690 nm

λ, nm	a^*_{chl}, m^2/mgr	a^*_{sm}, m^2/g	a^*_{doc}, m^2/gC	a_w, m^1	b^*_{bchl}, m^2/mg	b^*_{bsm}, m^2/g	b_w, m^{-1}
410	0.0380	0.2650	0.2800	0.0162	0.001250	0.0365	0.0052
430	0.0400	0.2300	0.2500	0.0144	0.001230	0.0250	0.0042
450	0.0410	0.2000	0.2300	0.0145	0.001210	0.0270	0.0035
470	0.0400	0.1800	0.1800	0.0156	0.001200	0.0290	0.0029
490	0.0340	0.1600	0.1600	0.0196	0.001210	0.0305	0.0024
510	0.0280	0.1450	0.1400	0.0357	0.001240	0.0320	0.0020
530	0.0220	0.1300	0.1250	0.0507	0.001270	0.0330	0.0017
550	0.0180	0.1200	0.1100	0.0638	0.001290	0.0335	0.0015
570	0.0150	0.1100	0.1000	0.0799	0.001280	0.0330	0.0013
590	0.0130	0.1050	0.0900	0.157	0.001270	0.0325	0.0011
610	0.0120	0.1000	0.0800	0.289	0.001270	0.0320	0.001
630	0.0120	0.1000	0.0700	0.319	0.001260	0.0310	0.0009
650	0.0200	0.1050	0.0600	0.349	0.001220	0.0290	0.0007
670	0.0250	0.1150	0.0500	0.43	0.001160	0.0270	0.0007
690	0.0160	0.1250	0.0500	0.5	0.001080	0.0250	0.0006

Table 16.2 The CIE color mixtures (for red, green and blue) for equal energy spectra

λ, nm	x"	y"	z"
410	0.0840	0.0023	0.4005
430	0.5667	0.0232	2.7663
450	0.6730	0.0761	3.5470
470	0.3935	0.1824	2.5895
490	0.0642	0.4162	0.9313
510	0.0187	1.0077	0.3160
530	0.3304	1.7243	0.0841
550	0.8670	1.9906	0.0174
570	1.5243	1.9041	0.0042
590	2.0535	1.5144	0.0023
610	2.0064	1.0066	0.0007
630	1.2876	0.5311	0.0000
650	0.5681	0.2143	0.0000
670	0.1755	0.0643	0.0000
690	0.0457	0.0165	0.0000

CIE commission Internationale de l'Éclairage

Table 16.3 Subsurface incident radiation E_d (−0,λ) for sun zenith angles 0°, 30°, 45°, 60°

λ, nm	E_d (−0,λ), W/m^2 nm			
	$\theta_0 = 0°$	$\theta_0 = 30°$	$\theta_0 = 45°$	$\theta_0 = 60°$
410	1.33	1.27	1.19	1.03
430	1.27	1.21	1.15	1.00
450	1.64	1.56	1.48	1.30
470	1.65	1.57	1.50	1.32
490	1.59	1.52	1.46	1.29
510	1.64	1.56	1.50	1.33
530	1.62	1.55	1.49	1.33
550	1.63	1.56	1.51	1.34
570	1.59	1.52	1.47	1.32
590	1.54	1.47	1.43	1.28
610	1.49	1.43	1.39	1.25
630	1.44	1.38	1.34	1.20
650	1.35	1.30	1.26	1.13
670	1.32	1.27	1.24	1.11
690	1.24	1.19	1.16	1.04

from the files embedded into the respective codes of each lab. The description of Practice 15 contains also some examples of results, which are expected to be reached through the fulfillment of this numerical exercise (Figs. 16.1 and 16.2 as well Tables 16.5 and 16.6).

Table 16.4 Spectral values of *F* for sun zenith angles 0°, 30°, 45°, 60°

λ, nm	F, relative units			
	$\theta_0 = 0°$	$\theta_0 = 30°$	$\theta_0 = 45°$	$\theta_0 = 60°$
410	0.32	0.38	0.42	0.55
430	0.30	0.35	0.38	0.50
450	0.27	0.33	0.35	0.46
470	0.25	0.31	0.32	0.42
490	0.23	0.29	0.30	0.39
510	0.21	0.28	0.28	0.37
530	0.20	0.26	0.26	0.35
550	0.19	0.25	0.25	0.33
570	0.18	0.23	0.24	0.32
590	0.17	0.22	0.23	0.31
610	0.17	0.21	0.22	0.29
630	0.16	0.20	0.21	0.28
650	0.15	0.20	0.20	0.27
670	0.14	0.19	0.19	0.26
690	0.14	0.18	0.18	0.25

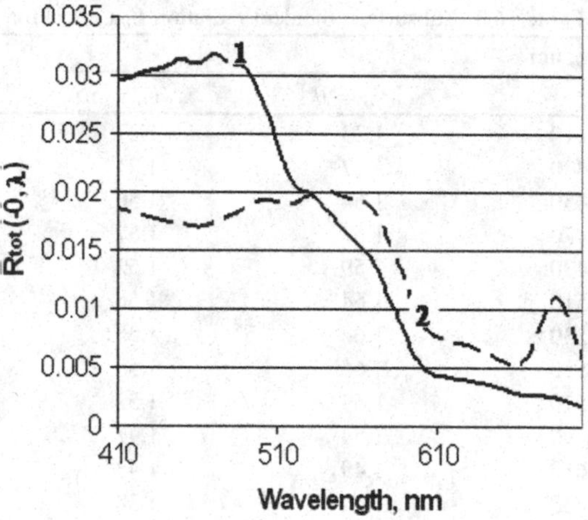

Fig. 16.1 Simulated total coefficient of water column diffuse reflection, $R_{tot}(-0,\lambda)$ for the input data given in Table 16.5

16.2 Mechanisms of Interactions of Solar Light with Absorbing and Scattering Aquatic Media

Of all components constituting the radiant signal recorded by the remote sensor overflying aquatic environments, it is only the radiance emerging from beneath the water surface, $L_w(+0, \lambda)$ carries the information on both the water quality and the nature of sun photon interactions with the aquatic medium. However, being simultaneously the function of optical properties of water per se and the intensity and

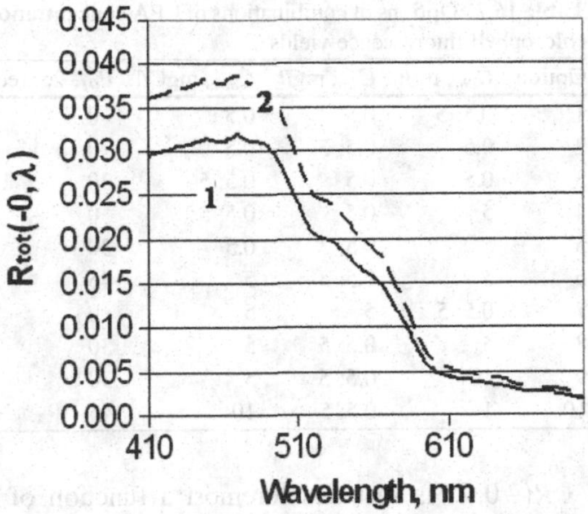

Fig. 16.2 Simulated total coefficient of water column diffuse reflection, R_{tot} (−0,λ) for the input data given in Table 16.6

Table 16.5 Input data for simulating the resulting coefficient of water column diffuse reflectance, R_{tot} (−0,λ)

C_{chl}, μg/l	C_{sm}, mg/l	C_{doc}, mgC/l	θ_0, degree	*h*, m	η_{chl}, rel. units, %	Bottom cover type	Curve in Fig. 16.1
0.1	0.1	0.1	30	100	3.0	Sand	1
3.0	0.1	0.1	30	100	3.0	Sand	2

Table 16.6 Input data for simulating the water column total coefficient of diffuse reflection, R_{tot}(−0,λ)

C_{chl}, μg/l	C_{sm}, mg/l	C_{doc}, mgC/l	θ_0, degree	*h*, m	η_{chl}, rel. units, %	Bottom cover type	Curve in Fig. 16.2
0.1	0.1	0.1	30	100	3.0	Sand	1
0.1	0.1	0.1	60	100	3.0	Sand	2

spectral composition of the incident solar flux, the upwelling radiance $L_w(+0, \lambda)$ proves to be inconvenient for further analysis of the influence exerted by the suit of various inherent light transfer processes on the water leaving signal. It is more convenient to use for this purpose the coefficient of light diffuse reflection in water just beneath the atmosphere-water interface, $R(-0, \lambda)$, which is intimately/functionally related to the upwelling radiance, $L_w(+0, \lambda)$.

By definition, $R(-0, \lambda)$ is the upwelling irradiance, $E_u(-0, \lambda)$ just beneath the water surface normalized to the downwelling irradiance, $E_d(-0, \lambda)$ at the same level/vertical distance from the interface (see Table 16.7):

$$R(-0, \lambda) = \frac{E_u(-0, \lambda)}{E_d(-0, \lambda)} \tag{16.1}$$

Table 16.7 Options of combinations of CPA concentrations, sun zenith angles, bottom depths and chlorophyll fluorescence yields

Option	C_{chl}, μg/l	C_{sm}, mg/l	C_{doc}, mgC/l	θ_{01}, degree	θ_{02}, degree	h_1, m	h_2,m	η_{chl}, r.u.
1	0.5; 5	0.5	0.5	0	30	1	5	0.3; 3.0
2	0.5	0.5; 3	0.5	0	45	1	10	0.1; 1.0
3	0.5	0.5	0.5; 5	30	45	5	10	0.2; 2.0
4	5	0.5	0.5; 5	0	60	1	20	1.0; 1.5
5	5	0.5; 3	0.5	30	60	5	20	0.5; 2.5
6	5	0.5; 3	5	45	60	10	20	2.0; 3.0
7	0.5; 5	5	5	0	30	1	20	1.0; 3.0
8	5	0.5; 5	5	30	45	1	10	1.0; 2.0
9	10	0.5; 5	5	30	60	1	5	0.3; 4.0
10	1	0.5; 5	10	45	60	1	5	1.0; 15.0

$R(-0,\lambda)$ is first and foremost a function of inherent hydro-optical properties (IOPs) incorporating generally speaking coefficients of absorption, a, scattering, b (including backscattering, b_b), direct light attenuation, c as well as the phase functions of elastic scattering. Due to the property of additivity, inherent optical properties, reflect in turn the composition of the aquatic medium, and hence, the individual contribution of each CPA to the formation of the upwelling/emerging light.

In the single scattering approximation, assuming that the light fields originating from sources of different nature can be considered as independent, the apparent hydro-optical characteristics ($R(-0,\lambda)$ is one of them) are additive functions of apparent hydro-optical characteristics arising from each type of light interaction with the aquatic medium that is absorption/elastic scattering in a semi-infinite water layer $R_{e.s.}$, water Raman scattering R_r, and fluorescence (R^f_{chl} и R^f_{doc}), as well as of the light interaction with the interfaces. Consequently

$$R(-0,\lambda) = R_{e.s.}(-0,\lambda) + R_r(-0,\lambda) + R^f_{chl}(-0,\lambda) + R^f_{doc}(-0,\lambda) + R_{bot}(-0,\lambda). \tag{16.2}$$

The natural water fluorophores are first of all chlorophyll and dissolved organic matter. Of all interface interactions, the light interaction with the bottom (R_{bot}) deserves particular attention in the context of the present teaching lab. The reader interested in deeper familiarization with the light interactions at the water-air interface is recommended to address dedicated monographs.

Earlier, a number of mathematical expressions had been suggested to relate $R_{e.s.}(-0,\lambda)$ with the IOPs. One of them was developed for a wide range of hydro-optical conditions (i.e. values of IOPs) which practically encompasses the variety of situations occurring in temperate surface and marine coastal waters and is valid for solar zenith angles not exceeding ~50°:

$$\begin{aligned} R_{e.s}(-0) &= (1/\mu_0)0.319 b_b/a \quad \text{for} \quad 0 \le b_b/a \le 0.25, \\ R_{e.s.}(-0) &= (1/\mu_0)[0.013 + 0.267 b_b/a] \quad \text{for} \quad 0.25 \le b_b/a \le 0.50, \end{aligned} \tag{16.3}$$

with $\mu_0 = \cos(\theta_0')$, θ_0' being the in-water refracted angle, backscattering coefficient $b_b = Bb$, B being the backscattering probability, $\mu_0 = 0.858$ for overcast conditions, (−0) indicating just beneath the water surface, a and b are respectively coefficients of absorption and scattering within the water column.

It is known that the IOPs of an aquatic medium (in our case the coefficients of adsorption and backscattering) are additive in nature and can be expressed as a sum of products of absorption (a*) and backscattering (b_b*) cross-sections and the concentrations of respective CPAs – C_i:

$$a = a_w + \sum_i C_i a_i^* \, ; \; b_b = (b_b)_w + \sum_j C_j (b_b)_j^*, \tag{16.4}$$

where subscripts w, i and j stand, respectively, for water and co-existing absorbing and scattering water medium components, the overall number of which does not necessarily need to be equal. Thus, given tabulated spectral values of absorption and backscattering cross-sections for each CPA, its individual contribution to the hydro-optical bulk properties of the target water column can be related to its concentration (Table 16.8). According to what was pointed out in the Introduction, we'll be using the *hydro-optical model* (tabulated spectral values of the coefficients of absorption and backscattering of phytoplankton chlorophyll, mineral suspended particulates, and dissolved organic matter) typical of Lake Ladoga.

Within the framework of Practice 15 we omit the influence of water Raman scattering on $R(-0,\lambda)$. It's noteworthy that Raman scattering affects noticeably the upwelling light only in clear, open ocean waters, this influence being pronounced in the long-wave part of the visible spectrum: R_r accounts for about 20% of the total value of $R(-0,\lambda)$. However, in coastal marine waters and inland waters, generally

Table 16.8 Spectral values of the bottom albedo, A (in relative units) for different types of bottom cover

λ, nm	Sand	Silt	Algae
410	0.27	0.0511	0.0271
430	0.32	0.0641	0.0299
450	0.34	0.0800	0.0300
470	0.35	0.0886	0.0377
490	0.37	0.0947	0.0461
510	0.39	0.1079	0.0543
530	0.41	0.1229	0.0613
550	0.44	0.1300	0.0700
570	0.47	0.1377	0.0940
590	0.49	0.1468	0.1108
610	0.51	0.1510	0.1025
630	0.53	0.1495	0.0884
650	0.56	0.1500	0.0800
670	0.58	0.1579	0.0502
690	0.60	0.1630	0.1013

rich in various CPAs, the actual influence of R_r on $R(-0,\lambda)$ proves to be insignificant.

Inclusion of fluorescence into the equation of light transfer, and the ensuing solution of this modified equation in the single scattering approximation for a vertically homogeneous aquatic media containing fluorescent agents (whose fluorescent yield is independent of λ and the emission line is Gaussian- shaped) gives the following expression for the volume reflectance R^f due to chlorophyll at the maximum wavelength of emission ($\lambda_{\max}$)

$$R^f_{chl} = \frac{\eta_{chl}}{\sqrt{2\pi\sigma^2}} \quad \exp\left(-\frac{(\lambda_{em}-\lambda_{0em})^2}{2\sigma^2}\right) / 2\, K_d(\lambda_{em})\ \mu_0 E_d(\lambda_{em},-0)\ \lambda_{em}$$
$$\times \int\limits_{\lambda_{ex}} \lambda_{ex}\ a_f(\lambda_{ex})\ E_d(\lambda_{ex},-0)\ f(\lambda_{ex},\lambda_{em})\ d\lambda_{ex} \tag{16.5}$$

where η is the fluorescence yield, λ_{ex} and λ_{em} are the excitation and emission wavelengths respectively, $\lambda_{o\ em}$ is the wavelength of fluorescence band center, $a_f(\lambda)$ is the fluorophore absorption coefficient, σ is the half-width of the fluorescence band with a Gaussian shape,

$$f(\lambda_{ex},\lambda_{em}) = \frac{K_d(\lambda_{em})}{K_d(\lambda_{ex})}\left[1-\frac{K_d(\lambda_{em})}{K_d(\lambda_{ex})}\ln\left(1+\frac{K_d(\lambda_{ex})}{K_d(\lambda_{em})}\right)\right],$$

$K_d(\lambda)$ is the downwelling irradiance attenuation coefficient, $E_d(\lambda,-0)$ is the incident irradiance just beneath the water surface.

It is known that the chlorophyll fluorescence band is centered at 685 nm, and its width at half-maximum is about 25 nm.

The volume reflectance coefficient R^f_{doc} arising from the DOC fluorescence can be numerically assessed from (16.6):

$$R^f_{doc}\ (\lambda_{em}) = \frac{1}{2\sqrt{2\pi\,\sigma^2}} \exp\left(-\frac{(\lambda_{em}-\lambda_{0em})^2}{2\sigma^2}\right) \times$$
$$\times \int\limits_{\lambda_{ex}} \eta_{doc}\,(\lambda_{ex})\, a_{doc}\,(\lambda_{ex})\, \frac{E_d(-0,\lambda_{ex})}{\mu_0 \quad (K_d(\lambda_{ex})+2\mu_0\ K_d(\lambda_{em}))}\, d\lambda_{ex}, \tag{16.6}$$

where λ_{doc} is the dissolved organic matter (DOM) fluorescence yield, a_{doc} is the fluorophore (DOM) absorption coefficient.

The maximum of the dissolved organic matter fluorescence band is located at ~ 490–520 nm, and its width at half-maximum is about 100 nm.

When deriving the relationship between $R(-0,\lambda)$ and the IOP's it was assumed that the aquatic medium is a *semi-infinite* absorbing/scattering layer. However, in the case of sufficiently transparent and shallow waters this assumption becomes invalid. Indeed, in such waters a certain amount of photons traveling downwards is reflected at the bottom back to the water-air interface instead of being absorbed by water molecules at depth. It is conducive to the probability that a certain fraction of reflected photons eventually reach the water surface. Thus, the resulting diffuse reflectance just beneath the water surface $R_{tot}(-0,\lambda)$ can be decomposed:

$$R_{tot}(-0,\lambda) = R_w(-0,\lambda) + R_{bot}(-0,\lambda), \tag{16.7}$$

where $R_w(-0,\lambda)$ is the upwelling spectral reflectance due to water (by definition it equals the ratio $E_u(-0,\lambda)/E_d(-0,\lambda)$, E_u, E_d, being upwelling and downwelling irradiances at z = −0) and $R_{bot}(-0,\lambda)$ is the upwelling spectral reflectance due to the bottom optical influence, which can be defined as:

$$R_{bot}(-0,\lambda) = (A - R_{e.s.}(-0,\lambda))\exp(-2K_d(\lambda)h), \tag{16.8}$$

A being the bottom spectral albedo.

Under cloudless conditions, the coefficient of downwelling irradiance attenuation can be parameterized as follows:

$$K_{sun}(-0,\lambda,\theta_0) = (1/\mu_0)\left[a^2(\lambda) + (0.473\mu_0 - 0.218)a(\lambda)b(\lambda)\right]^{1/2}, \tag{16.9}$$

where μ_0, as above, is equal to $\cos(\theta_0')$, θ_0' being the in-water refracted sun zenith angle.

For overcast conditions:

$$K_{sky}(-0,\lambda) = 1.168\left[a^2(\lambda) + 0.168a(\lambda)b(\lambda)\right]^{1/2}. \tag{16.10}$$

Hence, the attenuation coefficient of downwelling global radiation, $K_d(-0,\lambda,\theta_0)$, can be expressed as

$$K_d(-0,\lambda,\theta_0) = F_w K_{sky}(\lambda) + (1 - F_w)K_{sun}(\lambda,\theta_0), \tag{16.11}$$

where $F_w = F\left(1-\rho_{sky}\right)/\left[F\left(1-\rho_{sky}\right) + (1-F)(1-\rho_{sun}(\theta_0))\right]$, $\rho_{sky} = 0.066$ and $\rho_{sun}(\theta_0)$ are the Frésnel reflectivities of sky and solar irradiance (directly propagating from the zenith angle θ_0) respectively; $F = E_{sky}/\left(E_{sky} + E_{sun}\right)$ is the fraction of the incident irradiance that is diffuse. It is valid $\rho_{sky} = 0.066$, and the value $\rho_{sun}(\theta_0)$ can be numerically assessed from the Fresnel formulas:

$$\rho_{sun} = \frac{1}{2}\left(\frac{\sin^2(\theta_i - \theta_r)}{\sin^2(\theta_i + \theta_r)} + \frac{tg^2(\theta_i - \theta_r)}{tg^2(\theta_i + \theta_r)}\right) \tag{16.12}$$

where θ_i = incidence angle, θ_r = angle of refraction at the water-air interface. The angles θ_i and θ_r are interrelated through Snell's law:

$$\frac{\sin\theta_i}{\sin\theta_r} = n, \tag{16.13}$$

where n is the relative index of refraction of water, equal to 1.333. The values of F are tabulated in Table 16.4.

16.3 Practice 15

16.3.1 *Objectives*

1. Investigate the spectral composition of surface solar radiation diffuse reflectance just beneath the water surface, $R(-0,\lambda)$ as a function of solar zenith angle.
2. Investigate the spectral variations of $R(-0,\lambda)$ with the concentration of water colour producing agents (CPAs), such as phytoplankton chlorophyll, suspended mineral and dissolved organic matter.
3. Investigate the spectral variations of $R(-0,\lambda)$ with bottom depth and albedo.
4. Gain practical experience in the use and modernization of the already available software packages for conducting simulations at PCs.
5. Get acquainted with drawing up of the performed lab practice making use of the "Word" text editor with the emphasis of attaining experience in inclusion of tables, graphical and text files.

16.3.2 *Software and Set of Input Parameters*

1. "F_TASK" code in "*Paskal* v.7.0" (TP7), and files with the input data stored in the folder "c:\Dis_liq"
2. Text editor WORD, packages EXCEL, SURFER and TABLECURVE.
3. Set of input parameters provided by the teacher.

16.3.3 *Sequential Steps of the Exercise Implementation*

1. Read attentively the section devoted to the physical/theoretical background of this exercise. If necessary, consult the referenced literature.
2. Using the code "F_ TASK.exe" and the options suggested in Table 16.7, obtain and further analyze the spectral dependence of $R_{tot}(-0,\lambda)$ upon:

- sun zenith angles (within the given range of angles) (see Table 16.7) concentrations of chlorophyll, suspended matter, and dissolved organic carbon Table 16.7
- bottom depth
- a bottom albedo (i.e. bottom type) (see Table 16.8, which illustrates the spectral albedo of the bottom covered with three types of substances: sand, silt and grass); these data are automatically entered into the code from the data file as soon as the bottom type is selected.
- At each run of the program "F_TASK exe", the results of calculations are entered into to files "*reflect.dat*" and "*wv_i.dat*". The first of the files accommodates the values of the resulting coefficient of water column reflection, $R_{tot}(-0, \lambda)$, whereas the second file stores the respective wavelengths. Since at each program running, the data are reentered under one and the same name "*reflect.dat*", it should be given another name (e.g. "*reflect.dat*", "*reflect2.dat*" etc.), and write down in your pad of notes the parameters, with which these files were generated.

3. Prepare a concise report reflecting the principle stages of the numerical experiment, as well as the obtained results followed by the main conclusions. Based on the files containing the results of simulations and exploring the MS Excel facility, plot the function $R_{tot}(-0, \lambda)$, employing any suitable program package such as EXCEL, SURFER or else TABLECURVE.
4. Qualitatively explain the obtained dependencies based on the theory of light transfer in turbid and absorbing media (see the present manual and the recommended literature). In doing this, consider closely the spectral distributions of the absorption and backscattering cross-sections for the major CPAs.

16.3.4 Requirements to the Report

Compile a concise report reflecting the principal stages and the obtained results of the performed exercise in form of tables and plots (graphs). Importantly, the report should be concluded with the main corollaries. Several examples of tables and plots (graphs) that should be generated as a result of the exercise performance are given in Tables 16.5 and 16.6 as well as Figs. 16.1 and 16.2.

Chapter 17
Simulations and Analyses of Variations in Colorimetric Properties of Natural Waters with Specific Reference to Waters with Significant Spatial Heterogeneity of Optical Properties

Abstract Chapter 17 opens with a brief description of the way of quantification of water colour properties via such parameters as dominant wavelength and colour purity. This is done to introduce the reader to an approach of a more detailed perception of water colour formation under conditions of independently and in appreciably wide limits varying concentrations of the major colour producing agents, CPAs, such as phytoplankton, suspended minerals and dissolved organic matter. The exercises performed by the readers will permit them to realize that one and the same water colour might result from absolutely different combinations of the CPA concentration vector. It is hoped that this exercise will help the reader avoid very frequently occurring erroneous attributions of water colour as it is perceived from space over case II waters to some definite CPA, e.g. if the water colour is green, the water must be rich in phytoplankton, and if it is brown it should be abundant in terrigenous suspended matter with a predominant mineral component, etc. This exercise will also bring the reader to a clear vision why widely employed bio optical "band-ratio" algorithms can hold exclusively in case I waters and prove untenable in case of their application to retrieve CPAs in turbid marine coastal and inland waters.

17.1 Formation of Water Color: A Concise Description of the Physical/Theoretical Background

The human eye senses the water color through capturing and analyzing the radiance signal coming up from the water surface. Generally, this signal consists of two components $L_{surf}(+0,\lambda)$ and $L_u(+0,\lambda)$, originating respectively from the light both reflected by the water surface and scattered by the water column back into the atmosphere. This is the second component which carries information about colorimetric properties of the water column, and consequently about the OAC concentrations.

I. Melnikova et al., *Remote Sensing of the Environment and Radiation Transfer*,
DOI 10.1007/978-3-642-14899-6_17,

The relation between the radiance coming up from beneath the water surface $L_u(+0,\lambda)$ and the IOP's (coefficients of absorption a and backscattering b_b) could be parameterized via employing the volume reflectance quantity, R:

$$L_u(+0,\lambda) = \frac{R(-0,\lambda)\,(1-\rho(\theta))\,E_d(-0,\lambda)(1-\rho_{irr})}{\pi n^2(1-0.48\,R(-0,\lambda))} \quad (17.1)$$

where $R(-0,\lambda) = E_u(-0,\lambda)/\,E_d(-0,\lambda)$ is the volume reflectance just beneath the water surface, $E_u(-0,\lambda)$, $E_d(-0,\lambda)$ are the subsurface upwelling and downward subsurface irradiances respectively; $\rho(\theta)$ is the Fresnel reflectivity within the field-of-view of the remote sensing instrument at the nadir angle of viewing θ; n is the relative refraction index of water ($n = 1.333$); ρ_{irr} = surface reflectivity for downward irradiance in air.

For reasons of simplification, the angle θ will be fixed at 0°, which corresponds to a strictly vertical viewing of the water surface. In the exercise calculations, the term (1-ρ_{irr}) is also omitted as it accounts for a less than 1% difference in the value of Lu.

As it was stated above (see Chap. 16), the tabulated spectral values of absorption and backscattering cross sections (also called *hydrooptical model*) for Lake Ladoga will be used throughout the numerical simulations.

Within the framework of the chromaticity analysis, the upwelling radiance spectrum can be related to the color sensed by a human being through integrating the human eye's sensitivity and the upwelling light spectrum. The resulting tristimulus values are then given by:

$$\begin{aligned} X' &= \int x''(\lambda)L_u(+0,\lambda)d\lambda \\ Y' &= \int y''(\lambda)L_u(+0,\lambda)d\lambda \\ Z' &= \int z''(\lambda)L_u(+0,\lambda)d\lambda \end{aligned} \quad (17.2)$$

where x'', y'', z'' are the CIE (*Commission Internationale de l'Èclairage*) color mixtures (for red, green and blue respectively) for equal energy spectra and may be obtained from the CIE tables (see Table 16.2).

The chromaticity coordinates could then be obtained from the following equations:

$$\begin{aligned} x &= \frac{X'}{X'+Y'+Z'} \\ y &= \frac{Y'}{X'+Y'+Z'} \\ z &= \frac{\dot{Z}}{X'+Y'+Z'} \end{aligned} \quad (17.3)$$

So far as $x + y + z = 1$, two chromaticity coordinates adequately represent color in a chromaticity diagram and the chromaticities can be displayed as 2-D plots of either y (green) – z (blue) or x (red) – y (green).

Using the CIE color values, $x'(\lambda)$, $y'(\lambda)$, $z'(\lambda)$ and assuming monochromatic light of a given wavelength as the spectrum $E(\lambda)$, the CIE chromaticity coordinates may be obtained for that particular wavelength. By repeating this procedure, the CIE chromaticity coordinates may be obtained for each wavelength throughout the visible spectrum. All the (x, y) pairs thus obtained are plotted in Fig. 17.1, delineating the so called *color triangle*. For a white spectrum when

$$L(\lambda) = const,$$

and the color coordinates are mutually equal

$$x = y = z = 0.333.$$

This defines the achromatic color or white point **S** illustrated in Fig. 17.1.

A numerical value of color is then obtained by drawing a line from this white point **S** through the plotted chromaticity values of the measured spectrum (as indicated by point **Q**). The intersection of the line **S-Q** with the curve envelope of Fig. 17.1 (indicated by point **A**) specifies the *dominant wavelength* λ_{dom} that will herein be considered as the colorimetric definition of the natural water body.

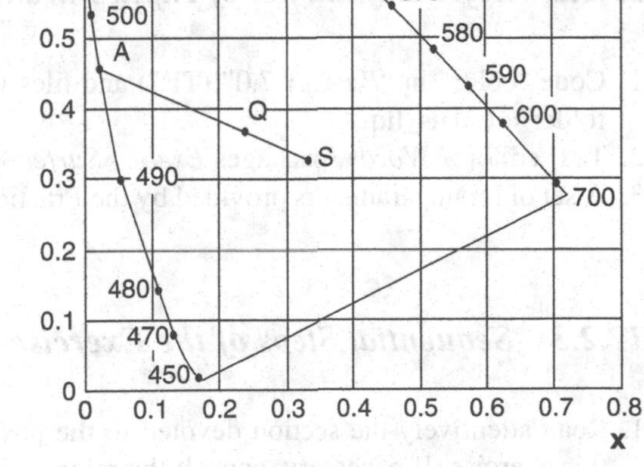

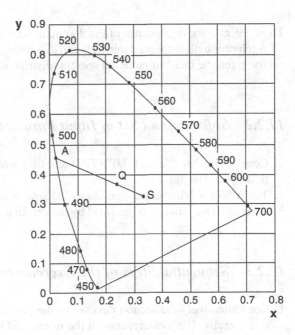

Fig. 17.1 A plot for determining the water radiometric characteristics

The distinctiveness of this dominant wavelength is termed "*color purity*" and is defined in Fig. 17.1 as the ratio of the line **Q-S** to the line **A-S**. Thus, spectral purity is a measure of the magnitude of the contribution of the dominant monochromatic spectrum at the dominant wavelength λ_{dom}, while a spectral purity p of 0 indicates a "white" spectrum. Together, the dominant wavelength λ_{dom}, and its associated spectral purity p are considered herein as defining aquatic color.

Since the upwelling radiance $L_u(+0,\lambda)$ is controlled by the CPA optical influence (transduced through $R(-0,\lambda)$, see Practice 15), any changes in chlorophyll, suspended minerals and dissolved organic carbon are bound to result in changes of the water column color. In the case of significant horizontal heterogeneity of the hydro-optical field structure of the target water body, the color must also display a distinct patchy pattern. Such a phenomenon is characteristic of many lakes and water storage reservoirs, as well as marine coastal zones. It can be easily detected on space imageries in the visible spectrum.

Focused on studying the formation of radiometric characteristics of natural water, this teaching lab is confined to a simplified numerical modeling experiment when chlorophyll and dissolved organic carbon fluorescence impacts are neglected. A complete solution of this problem can be found elsewhere.

17.2 Practice 16

17.2.1 Objective

To investigate the dependence of the dominant wavelength and color purity on the CPA (chlorophyll, suspended minerals and dissolved organic carbon) concentration vector given the incident radiation spectral distribution.

17.2.2 Software and Set of Input Parameters

1. Code "color" in "*Paskalv*.7.0" (TP7) and files with the input data stored in the folder « c:\Dis_liq ».
2. Text editor « *Word*», packages *Exel*», «*Surfer* » and « *TableCurve* ».
3. A set of input parameters provided by the Practice chief/supervisor. (Table 17.1)

17.2.3 Sequential Steps of the Exercise Implementation

1. Read attentively the section devoted to the physical/theoretical background of this exercise. If necessary, consult the referenced literature.

Table 17.1 Recommended combinations of CPA concentrations[a]

Option	C_{chl}, μg/l	C_{sm}, mg/l	C_{doc}, mgC/l
1	0.1; 1; 5; 10	0	1
2	0.1; 1; 5; 10	0	5
3	0.1; 1; 5; 10	0	10
4	0	0.1; 1; 5; 10	1
5	0	0.1; 1; 5; 10	5
6	0	0.1; 1; 5; 10	10
7	1	1	0.1; 1; 5; 10
8	5	1	0.1; 1; 5; 10
9	1	5	0.1; 1; 5; 10
10	5	5	0.1; 1; 5; 10

[a]As $E_d(-0, \lambda)$, any spectral distribution of incident radiation can be taken from Table 16.3 in dependence of the sun zenith angle

Table 17.2 Input data and results of simulations of color coordinates x, y as well as dominant wavelength, λ_{dom}, and color purity, p of the water column

C_{chl}, μg/l	C_{sm}, mg/l	C_{doc}, mgC/l	θ_0, deg.	x	y	λ_{dom}, nm	p
0.1	0.1	0.1	30	0.208	0.261	484	0.4939
0.1	0.0	5.0	30	0.278	0.315	487	0.2033
0.1	0.1	5.0	30	0.334	0.370	556	0.1168

2. Using the code "color" and combinations of the CPA concentrations recommended in Table 17.1, perform the simulations of water color characteristics and analyze:

   – the dominant wavelength dependence on a number of the CPA combinations;
   – the color purity dependence on the same as above CPA combinations.

Note: The code "color" permits to display the simulation results on the PC screen: you only need to push simultaneously two buttons "*Ctrl+O*".

17.2.4 Requirements to the Report

– Using the text editor "World", make (draw) a concise report reflecting the major steps of the implemented Exercise. Based upon the basic theoretical considerations, the results obtained should be presented as a table and further discussed in terms of the established regularities in the water color formation under different conditions of CPA concentrations. The report should be closed with the main corollaries/conclusions.
– Table 17.2 is an example of the table you need to generate as a result of the teaching lab performance. The report should be stored on the network drive.

Chapter 18
Retrieval of CPA Concentrations from the Spectral Composition of Subsurface Water Column Diffuse Reflectance: Application to Environmental Remote Sensing Tasks

Abstract Chapter 18 opens with a concise review of physical and methodological approaches addressing the retrieval of CPAs in waters of different optical complexity. It is explained why the determination of phytoplankton chlorophyll concentration in clear waters can be affected via using relatively simple two or three band-ratio algorithms based on statistically ample in situ data and frequently involving a semi-analytical expressions. However, much more sophisticated methods (illustrated in Chapter 18 by the Levenberg-Marquardt technique as an example) are required to remotely sense phytoplankton chlorophyll in turbid/strongly absorbing waters: in such cases, it is mandatory to simultaneously determine the concentrations of all coexisting CPAs. Only then the concentration of chlorophyll can be accurately determined. Besides, the simultaneously retrieved concentrations of other main CPAs are important for interpreting the ecological state of the waters under investigation.

The reader is offered a set of combinations of CPAs (i.e. hydro-optical situations, all of which are subsumed under the category of case II waters) for which the exercises should be performed. The Practice no. 17 is designed in such a way that the reader have to creatively apply the respective computer program in order to attain the CPA retrievals at the highest possible accuracy.

18.1 Methods of Retrieval of Water Quality from Remotely Sensed Data in the Visible

When remotely studying natural water bodies, it is the above water upwelling radiance, $L_u(+0,\lambda)$ that is exploited for this purpose: the desired information about the content of water constituents is contained in the *spectral* composition of, $L_u(+0,\lambda)$. However, for the reasons given in Chap. 16, the subsurface volume reflectance $R(-0,\lambda)$ is a more convenient quantity for attaining this goal. The water surface radiance can be related to the optical characteristics of the water column through the volume reflectance $R(-0,\lambda)$ which in turn is a function of such IOP's as

I. Melnikova et al., *Remote Sensing of the Environment and Radiation Transfer*,
DOI 10.1007/978-3-642-14899-6_18,

the absorption and backscattering coefficients, a, b_b (see (16.3)). Due to their additive nature, the absorption and backscattering coefficients are sums of products of cross sections a^* and b_b^* (see (16.4)) and the respective CPA concentrations. As it was indicated in Chap. 16, the tabulated values of spectral cross-sections (or otherwise the *hydro-optical model*) determined for Lake Ladoga will be used herein for the simulations.

When processing remote sensing data collected over open ocean/sea waters, most frequently used are the methods based on regression expressions relating the chlorophyll concentration to the water-leaving radiance ratio at two wavelengths in the blue and green spectral regions.

As well known, the chlorophyll absorption spectrum exhibits two major absorption bands at 430–450 and 660–680 nm. When a water body containing phytoplankton (referred to as chlorophyll) is illuminated by the natural light, water-leaving radiance proves to be subdued (as compared to the radiance signal leaving *clear* water, i.e. water devoid of chlorophyll) in the region 400–500 nm (λ_1) and enhanced at $\lambda > 580$–600 nm. These changes result from, respectively, chlorophyll absorption and an increase in the number of scattering centers due to phytoplankton cells. At the same time, in the upwelling radiance spectrum there is a region ~ 500–520 nm (λ_2) where the signal remains nearly intact with increasing chlorophyll concentration. In this specific situation, there is a possibility to relate the normalized depth of the "chlorophyll dip" (i.e. $L_u(+0)$ at 430 nm) to the chlorophyll concentration:

$$Cchl = \alpha \left(\frac{L_u(\lambda_1 = 430\,nm)}{L_u(\lambda_2 = 520\,nm)} \right)^{-\beta}, \tag{18.1}$$

where α and β are regression coefficients obtained from statistically ample data incorporating concurrent *in situ* measurements of C_{chl} and remotely sensed L_u(420 nm) and L_u(520 nm).

A small but infinitesimal concentrations of dissolved organic carbon and suspended minerals, the increase in chlorophyll content results in a displacement of λ_1 and λ_2 to longer wavelengths [to 450–480 nm and 520, respectively]. In these conditions, the above "band-ratio" approach remains applicable: various λ_1/λ_2 pairs (443/520), (443/550), (520/550), (520/670) could be tried for a given combination of C_{chl}, C_{sm}, C_{doc} to fit best Eq. 18.1. Indeed, such attempts proved to be reasonably successful when dealing with clear off-shore ocean and marine waters. However, as soon as suspended minerals and dissolved organics concentrations sensibly increase (which is a characteristic of inland and marine coastal waters), it becomes impossible to identify λ_1 and λ_2 in the spectral distribution of the upwelling radiance, and the regression retrieval algorithm becomes completely inefficient.

One of the modern algorithms for water quality retrieval is the method of neural networks. It is based on approximation of the dependency of $R(-0,\lambda)$ on combinations of CPA concentrations. In the first stage, a training array of data containing the CPA concentration vector and the respective values of $R(-0,\lambda)$ at a number of wavelengths is developed. In the second stage, the neural network training is performed: the weighting coefficients are established in a way that the sum of squared residuals gets minimized. Due to its fast operation, this method provides for processing space imageries of vast water areas. In more details the theory and application of this method are given in the referenced publications.

Within the framework of this Exercise, consider more closely another method of water quality retrieval from remote sensing data in the visible, *viz*, method of multivariate optimization (the Levenberg-Marquardt method). As it was indicated above, the tabulated spectral values of CPA cross-sections for Lake Ladoga will be used herein for simulations.

If $R(\lambda, \boldsymbol{C}, a, b_b)$ is the water volume diffuse reflectance calculated using a known parameterization (e.g. (16.3)), and $\{S_j\}$ is the value of water volume diffuse reflectance obtained from *in situ* measurements, then the weighted residuals can be taken as a measure of concordance between the measured and simulated volume reflectance:

$$g_j = [S_j - R(\lambda, C, a, b_b)]/S_j, \tag{18.2}$$

where j is the number of wavelengths at which the measurements have been run.

Within the framework of the least squares method, the value of the concentration vector $\boldsymbol{C}$ (C_{chl}, C_{sm}, C_{doc}) can be found through minimizing the function of residuals over $\boldsymbol{C}$, $f(\boldsymbol{C})$:

$$f(C) = \sum\nolimits_j g_j^2(c), \tag{18.3}$$

Iterative calculations of the $f(\boldsymbol{C})$ minimum can be conducted using Levenber-Marquardt method, which, being a variant of the Newton-Gauss method, is more easily converging. To find an optimal concentration vector, the following iteration formula is used:

$$C_{k+1} = C_k + \lambda_k \left(F_k^t F_k + \mu_k D_k\right)^{-1} F_k^t \left(1 - \frac{R(C_k)}{S_k}\right), \tag{18.4}$$

where $D_k = diag\left(F_k^t F_k\right)$ is a diagonal matrix, the main diagonal of which is composed of the elements $F_k^t F_k$, $F(C) = \left\|\frac{\partial R_i}{\partial C_j}\right\|$ is a matrix of the $n \times m$ order (n is the number of wavelengths, m is the dimension of the concentration vector $\boldsymbol{C}$), $F^t(C)$ is a transposed matrix $F(C)$, μ_k is the direction of minimization ($\mu_k \xrightarrow[k \to \infty]{} 0$), λ_k is the step of minimization, which is chosen based on the condition:

$$f(C_k + \lambda P_k) - f(C_k) < -\tau\lambda(q_k P_k), \tag{18.5}$$

where $P_k = \left(F_k^t F_k + \mu_k D_k\right)^{-1} F_k^t \left(1 - \frac{R(C_k)}{S_k}\right)$, $q_k = 2F_k^t \left(1 - \frac{R(C_k)}{S_k}\right)$.

The method of reduction is the simplest and most efficient method of choosing the step. This method consists of the following: choose two constants $0 < \tau < 1$ and

$0 < \chi < 1$, and set $\lambda = 1$ and test inequality (18.5). If inequality (18.5) holds, assume that $\lambda_k = \lambda$, and $a_{k+1} = a_k + \lambda_k P_k$. In the opposite case, reduce the pace by χ times, assuming $\lambda = \chi\lambda$, and test again inequality (18.5). This iteration procedure is repeated till inequality (18.5) is fulfilled.

The search for the desired minimum is only successful when the starting value of the concentration vector is close to the quaesitum. If the starting value of $\boldsymbol{C_0}$ proves to be far from the quaesitum, the minimum found for $f(\boldsymbol{C})$ might correspond to some unrealistic values of $\boldsymbol{C}$.

To avoid such difficulty, a number of starting values of $\boldsymbol{C_0}$ can be chosen, and making use of the method of multivariate optimization, a search for the most deep minimum is conducted. Yet, there is no guarantee that any particular starting point C_0 will result in the iterative procedure convergence or else the concentration vector will prove to be physically sound (e.g. negative concentrations of one or several CPAs). To obviate such difficulties, the following constrain is imposed on $\boldsymbol{C_0}$ such that

$$C_{i\,min} \le C_i \le C_{i\,max}, \tag{18.6}$$

18.2 Practice 17

18.2.1 Objectives

1. Explore one of the algorithms developed for retrieval of concentrations of such water quality constituents/CPAs as chlorophyll, suspended minerals and dissolved organic carbon from the spectral values of subsurface volume reflectance $R(-0,\lambda)$
2. Investigate the accuracy of retrieval of chlorophyll, suspended minerals and dissolved organic carbon depending upon both the concrete combinations of concentrations of these components and the initial concentration vectors in the iteration cycle

18.2.2 Software and Set of Input Parameters

1. Code “LM.exe” in *Paskal v. 7.0″* (*TP7*) and files with the input data stored in directory *\Dis_liq*”.
2. Text editor *WORD*, software packages EXCEL, SURFER or TABLECURVE.
3. Set of input parameters taken from the Table 18.1.

Table 18.1 Options of CPA concentration combinations

Option	C_{chl}, μg/l	C_{sm}, mg/l	C_{doc}, mgC/l
1	0.5; 1; 5; 18	0.5	1
2	0.5; 1; 5; 18	0.5	5
3	0.5; 1; 5; 18	0.5	18
4	0.5	0.5; 1; 5; 18	1
5	0.5	0.5; 1; 5; 18	5
6	0.5	0.5; 1; 5; 18	18
7	1	1	0.5;1; 5; 18
8	5	1	0.5; 1; 5; 18
9	1	5	0.5; 1; 5; 18
18	5	5	0.5; 1; 5; 18

Table 18.2 Input data and results of retrieval from $R(-0,\lambda)$ of the concentration of chlorophyll, suspended matter and dissolved organics

Input concentration vector			Initial concentration vector			Retrieval results			Mean root-square error
C_{chl}, μg/l	C_{sm}, mg/l	C_{doc}, mgC/l	C_{chl}, μg/l	C_{sm}, mg/l	C_{doc}, mgC/l	C_{chl}, μg/l	C_{sm}, mg/l	C_{doc}, mgC/l	-
3.0	3.0	3.0	1	1	1	3.37	3.34	3.78	0.0273
			2	2	2	3.29	3.44	3.61	0.0881

18.2.3 Sequential Steps of the Exercise Implementation

1. Having chosen several starting CPA concentration vectors for different values of the mean-square error, carry out the retrieval of the OAC concentration vector making use of the code "lm"; choose the most accurate retrieval results. As it ensues from above, the mean-square error according to the Eq. 18.4 is a criterion for exiting the iterative procedure of the concentration vector calculation with the Eq. 18.3. The closer its value to zero, the more accurate is the concentration vector retrieval. However, for different starting concentration vectors (which are necessary for initiating the iterative procedure) the mean-square error value close to zero can either be unattainable or hardly achievable because of unreasonably long computational time required for that. It is difficult, if ever possible, to specify/set up in advance the actual closeness of the mean-square error to zero for any initial (starting) concentration vector. That's why the students are recommended to first set up a tentative mean-square error value for each initial concentration vector and then consecutively decrease it after termination of each run of the code "IM.pas." Each step of this trail procedure should then be followed by assessment of the accurateness of the concentration vector retrieval.
2. Analyze the retrieval accuracy variations for each of the CPA depending on the concentrations of the rest CPAs given that the mean root-square error is fixed.

3. Prepare a brief report reflecting both the main stages of the numerical simulation experiment and the obtained results in the form of tables and figures followed by conclusions.

18.2.4 Requirements to the Report

Compile a concise report reflecting the main stages of the performed exercise, as well as the obtained results in form of tables and plots. The report should be closed with the main conclusions. An example of the table that should be generated as a result of the exercise implementation in given in Table 18.2.

References

Bohren CF, Huffman RK (1983) Absorption and scattering of light by small particles. Wiley, New York

Bukata RP, Jerome JH, Ya Kondratyev K, Pozdnyakov DV (1995) Optical properties and remote sensing of coastal and inland waters. CRC Press, Boca Raton, p 350

Charlson RJ, Heitzenberg J (eds) (1995) Aerosol forcing of climate. Wiley, Chichester

Cox CS, Munk WH (1954) Statistics of the sea surface derived from sun glitter. J Mar Res 13(2):198–227

Deirmendjian D (1969) Electromagnetic scattering on spherical polydispersions. Elsevier, New York

Goody RM, Yung UL (1996) Atmospheric radiation: theoretical basis. Oxford University Press, N.Y.

Harshvardhan, King MD (1993) Comparative accuracy of diffuse radiative properties computed using selected multiple scattering approximations. J Atmos Sci 50:247–259

Henyey L, Greenstain J (1941) Diffuse radiation in galaxy. Astrophys J 93:70–83

Hobbs PV (ed) (1993) Aerosol-cloud-climate interactions. Academic, San Diego

Joseph JH, Wiscombe WJ, Weiman JA (1976) The delta-Eddington approximation for radiative irradiance transfer. J Atmos Sci 33:2452–2459

Kargin BA (1984) Statistical modeling for solar radiation field in the atmosphere (in Russian). Printing House of SO AN USSR, Novosibirsk

Kolmogorov AN, Fomin SV (1989) Elements of the function theory and the functional analysis (in Russian). Nauka, Moscow

Korn GA, Korn TM (2000) Mathematical handbook for scientists and engineers: definitions, theorems, and formulas for reference and review, 2nd edn. Dower, New York

Lenoble J (ed) (1985) Radiative transfer in scattering and absorbing atmospheres: standard computational procedures. A. Deepak Publishing, Hampton

Makarova EA, Kharitonov AV, Kazachevskaya TV (1991) Solar irradiance (in Russian). Nauka, Moscow

Matveev LT (1965/2000) The foundations of general meteorology. Atmospheric Physics, Leningrad; St. Petersburg (in Russian) Gydrometeoizdat, 751 pp

Marchuk GI (ed) (1988) The Monte-Carlo method in the atmospheric optics (in Russian). Nauka, Novosibirsk

Melnikova I, Vasilyev A (2004) Short-wave solar radiation in the earth atmosphere: calculation, Observation, Interpretation. Springer/GmbH&Co. KG, Berlin/Heidelberg, p 310

Minin IN (1988) The theory of radiation transfer in the planets atmospheres (in Russian). Nauka, Moscow

I. Melnikova et al., *Remote Sensing of the Environment and Radiation Transfer*,
DOI 10.1007/978-3-642-14899-6,

Sivukhin DV (1980) The introductory survey of physics (in Russian). Nauka, Optics Moscow
Sobolev VV (1972) The light scattering in the planets atmospheres (in Russian). Nauka, Moscow
Stepanov NN (1948) The spherical trigonometry (in Russian). Moscow-Leningrad, Gostehizdat
Tikhonov AN, Ya Arsenin V (1986) Methods for solution of incorrect problems. Nauka, Moscow, p 288, in Russian
Van de Hulst HC (1957) Light scattering by a small particles. Wiley, New York
Van de Hulst HC (1980) Multiple light scattering: tables, formulas and applications, vol 1 and 2. Academic, New York
Wallace JM, Hobbs PV (1977) Atmospheric science (an introductory survey). Academic, Orlando
Ya Kondratyev K (1965) The actinometry (in Russian). Gydrometeoizdat, Leningrad
Ya Kondratyev K et al (1999) Limnology and remote sensing: a contemporary approach. Springer-Praxis, Chichester, 405 p
Yanovitskij EG (1997) Light scattering in inhomogeneous atmospheres. Springer, New York
Yu Sedunov S, Avdyushin SI, Borisenkov EP, Volkovitskiy OA, Petrov NN, Reintbakh RG, Smirnov VI, Chernov AA (eds) (1991) The atmosphere. Reference book (in Russian). Gydrometeoizdat, Leningrad
Yu Timofeev M, Vasilyev AV (2003) Theoretical principles of the atmospheric optics. Nauka, St. Petersburg, p 474, in Russian
Zuev VE, Belov VV, Veretennikov VV (1997) The theory of linear systems in the optics of diffuse medium (in Russian). Printing House of Siberian Filial RAS, Tomsk

Index

I. Melnikova et al., *Remote Sensing of the Environment and Radiation Transfer*,
DOI 10.1007/978-3-642-14899-6, © Springer-Verlag Berlin Heidelberg 2012